Praise for *A Slav*

A CONNECTICUT BOOK AWARD WINNER FOR NONFICTION
A CHRISTOPHER AWARD WINNER
BOOK SENSE PICK—DECEMBER 2007

"[A] splendid interpretation of the meaning of these men's experiences . . . an inestimable service to historians."—*New York Review of Books*

"The slave narratives of two Americans serve as eye-opening corridors to history."—*Washington Post*

"Two recently uncovered slave narratives create the backbone for this enthralling, intimate read."—*Chicago Tribune*

"[R]iveting... Breathtaking."—*New York Post*

"Perhaps the most moving aspect of the project is the way the men emerge as fully formed individuals."—*USA Today*

"[T]his testimony of overwhelming desire for freedom make[s] the book a powerful addition to the black history library."
—*Hartford Courant*

"Wonderful . . . a heart-rending and yet joyous book."
—*Charlotte News & Observer*

"The literature on American slavery is vast, and *A Slave No More* is an impressive and valuable addition to that canon."
—*Richmond Times-Dispatch*

"Remarkable . . . a fast-paced, intriguing, and original work, an historical detective work par excellence."—*Charlotte Observer*

"This instantly readable book seems destined for college reading lists and bedside tables because it offers compelling glimpses into slavery, and because of its powerful, if rough, prose."—*Cleveland Plain Dealer*

"[*A Slave No More*] shows in striking detail that slaves were not just passive recipients of freedom, but demanded and won it."—*America*

"A remarkable look at two lives."—*Yale Alumni Magazine*

"[T]his highly readable book tells two fascinating and profoundly moving stories."—*Historical Novel Review*

"Required reading for scholars or even casual students . . . essential for any collection on slavery, emancipation, or African American or U.S. history and literature."—*Library Journal*, starred review

"These powerful memoirs reveal poignant, heroic, painful, and inspiring lives."—*Publishers Weekly*

"A powerful, welcome addition to the Civil War library."
—*Kirkus Reviews*, starred review

"[*A Slave No More*] presents two of the most significant finds in the entire genre of slave narratives and of the primary material from the Civil War."
—DAVID LEVERING LEWIS, author of a
Pulitzer Prize-winning biography
of W.E.B. Du Bois.

"A compelling account of two men of remarkable courage who, by writing down their stories, sought to make themselves visible. Neither man could have wished for a more sympathetic or knowledgeable interpreter than David Blight."—CARYL PHILLIPS

"Together, Blight's meticulous research and the previously unknown autobiographical writings of these two men bring to life with unprecedented power the human dimensions of slavery and emancipation."
—ERIC FONER

A Slave No More

ALSO BY DAVID W. BLIGHT

Race and Reunion:
The Civil War in American Memory

Narrative of the Life of Frederick Douglass,
an American Slave, Written by Himself (editor)

Passages to Freedom: The Underground Railroad
in History and Memory (editor)

Beyond the Battlefield:
Race, Memory, and the American Civil War

Frederick Douglass' Civil War:
Keeping Faith in Jubilee

The Souls of Black Folk,
by W.E.B. Du Bois (editor)

A Slave No More

Two Men Who Escaped to Freedom
Including Their Own
Narratives of Emancipation

David W. Blight

AMISTAD

An Imprint of HarperCollinsPublishers

First Mariner Books edition 2009
Copyright © 2007 by David W. Blight

www.harpercollins.com

Library of Congress Cataloging-in-Publication Data
Blight, David W.
A slave no more: two men who escaped to freedom, including
their own narratives of emancipation/David W. Blight.—1st ed.
p. cm.
Includes first-person narratives of fugitive slaves
John Washington and Wallace Turnage.
Includes bibliographical references and index.
1. Washington, John, 1838–1918. 2. Turnage, Wallace, 1846–1916.
3. Fugitive slaves—United States—Biography. 4. Slaves—Virginia—
Fredericksburg—Biography. 5. Slaves—North Carolina—Green
County—Biography. 6. United States—History—Civil War,
1861–1865—African Americans. 7. African Americans—Biography.
8. Working class—United States—Biography.
9. Slave narratives—United States. I. Title.
E450.W325B58 2007
973.7'115—dc22 2007014467
ISBN 978-0-15-101232-9
ISBN 978-0-15-603451-7 (pbk.)

Text set in Adobe Caslon
Designed by Cathy Riggs

ScoutAutomatedPrintCode

To the memory of Martha A. Blight

By the rivers of Babylon, there we sat down,
yea, we wept, when we remembered Zion.
We hanged our harps upon the willows
in the midst thereof.
For there they that carried us away captive
required of us a song; and they that wasted
us required of us mirth, saying, sing us one
of the songs of Zion . . .
O daughter of Babylon, who art to be destroyed;
happy shall he be, that rewardeth thee as thou
hast served us.

—Psalm 137

Before morning I had begun to feel like I had truly
escaped from the hand of the slave master . . . I never
would be a slave no more.
—John Washington, remembering his first night
   of freedom along the Rappahannock River

I now dreaded the gun and handcuffs and pistols
no more. Nor the blowing of horns and the running
of hounds, nor the threats of death from the rebel's
authority.
—Wallace Turnage, remembering his first hours
   of freedom on a sand island in Mobile Bay

Contents

Contents

A Slave No More

Prologue

> No man can tell the intense agony which is felt by the
> slave, when wavering on the point of making his escape.
> All that he has is at stake; and even that which he has not
> is at stake also. The life which he has may be lost, and the
> liberty which he seeks, may not be gained.
> —FREDERICK DOUGLASS, 1855

John Washington, a twenty-four-year-old urban slave in Fredericksburg, Virginia, escaped across the Rappahannock River to Union army lines in April 1862 by ingenuity, skillful deception, and courage. Through the chaos of war he found his way to a tenuous freedom in Washington, D.C. Wallace Turnage, a seventeen-year-old slave born in North Carolina, ran away four times from an Alabama cotton plantation before he fled a Mobile slave jail on his fifth and final escape. In a harrowing and dramatic journey, he made his way on foot through swamps and snake-infested rivers to the safety of the Union navy in Mobile Bay in August 1864.

Both Washington and Turnage wrote remarkable post-emancipation narratives of their flights to freedom. Both also lived remarkable post-emancipation lives as laborers and citizens in Washington and New York City. In publishing their narratives and reconstructing their life stories for the first time, this book illuminates anew one of the greatest dramas of the Civil War—the anguished and glorious liberation of four million American slaves from generations of bondage.

———

Authors often spend years planning, researching, and outlining their books. But this book found me in a rather unusual way. In the spring of 2003 I was contacted by literary agent Wendy Strothman, who was working on behalf of Julian Houston—a writer and retired judge in Boston who had inherited John Washington's writings and papers from his mother, Alice Jackson Stuart. Alice Stuart had been the best friend of Washington's granddaughter Evelyn Easterly, who had saved and passed on her forebearer's writings. Wendy knew of my work on slave narratives, which have emerged in the past three decades as the founding texts of African American literature and an essential form of documentation for American slavery. Several years ago I edited a new edition of the most famous slave narrative of all, *Narrative of the Life of Frederick Douglass, an American Slave, Written by Himself.* Wendy sent me a photocopy of Washington's narrative and asked me to read it with an eye for authenticity and significance.

Months passed before I paid close attention, since I was preoccupied with other work and in the midst of a career move. In the fall of that year Debra Mecky, director of the Historical Society of the Town of Greenwich, Connecticut, invited me to give a lecture to her board of trustees. At dinner before the lecture, Debra asked me if I would be willing to have a look at what she and her staff believed to be an authentic slave narrative. It recently had been donated by Gladys Watts, a surviving friend of Wallace Turnage's daughter, Lydia Turnage Connolly, who had preserved the narrative in a special black clamshell box until her death in

Greenwich in 1984. My initial response was, "Another one?" Residents of the Town of Greenwich had owned small numbers of slaves in the late eighteenth and early nineteenth centuries. And now this wealthy small town had a former slave's dramatic story housed in their official heritage attic.

When I finally sat down and carefully read the two narratives together, I realized with a thrill of discovery that what we possessed were two extraordinary, unpublished, and probably unmediated narratives about one of the most revolutionary transitions in American history. If the lives of these two men could be verified, we would have two original ways of seeing how American slaves achieved freedom in the Civil War. Post-emancipation narratives by ex-slaves that told the story of their liberation from bondage in the 1860s are rare enough; in my lifetime a mere handful have surfaced, making these two untouched manuscripts very special.

Judge Houston had placed the Washington narrative on deposit at the Massachusetts Historical Society in Boston for professional safekeeping. Wendy Strothman arranged for me to meet Houston, and through the good services of the MHS, I saw the original manuscript of the Washington narrative and its many accompanying documents and photographs. We also went to Greenwich to see the original Turnage manuscript. Then we arranged an agreement among all the parties that I would combine the two narratives in one volume and tell the stories of Washington and Turnage together, including whatever I could uncover about their postwar lives.

In the case of Turnage's narrative, the HSTG had the wisdom to engage two people to work on the document before I was involved. Candis LaPrade, a literature professor at Tulane University, read the memoir and wrote a short essay placing it within the

genre of slave narratives. Most important, the HSTG hired
Christine McKay, an archivist at the New York Public Library's
Schomburg Center for Research in Black Culture, to investigate
Turnage's life after slavery. I in turn asked Christine to help me
do the same for the postslavery life of John Washington. Her
dogged pursuit of the details of each man's life has been crucial to
this book. With Christine's help, I have traced the post-
emancipation lives of both Washington and Turnage.

Their stories are both unique and representative. They are individual tales of freedom as well as two ways of seeing the whole
epic of emancipation. They remind us that history is unpre-
dictable, anguished, and hidden, but also sometimes patterned,
triumphant, and visible in the quiet and turbulent corners of the
lives of real people.

When political disunion came and the war broke out in 1861,
Northern free blacks and Southern slaves alike understood with
varying degrees of fear and hope that their fate lay in the balance.
They could imagine that the long-prophesied break with a dark
past was at hand. "At last our proud Republic is overtaken," wrote
the greatest black leader, Frederick Douglass, immediately after
the firing on Fort Sumter in April 1861. "Now is the time to
change the cry of vengeance long sent up from the ... toiling
bondsmen, into a grateful prayer for the peace and safety of the
government." Despite the untold "desolations" the war might
bring, Douglass now envisioned "armed abolition" conducted in a
sanctioned war against the South. Above all, he saw the war as a
crisis that "bound up the fate of the Republic and the fate of the

slave in the same bundle."[1] Such was the hope of 1861 for African Americans and their allies.

But most Americans—in the Confederate South as well as in the Union North—did not see the war in Douglass's way. The scope, character, and purpose of the war were yet to be determined. Was it time for Southern independence and the creation of a new Confederate slaveholding nation? Or was it time for the rapid restoration of the Union by suppressing the insurrection with force and returning to the prewar situation of coexisting slave and free states? Or, as blacks and their white abolitionist friends hoped, was it time for a revolution, a long war that would destroy slavery and reinvent a second American republic more inclusive than the first? What time was it? Throughout the war, in thousands of different circumstances, under changing policies and redefinitions of their status, and in the face of social chaos caused by huge military campaigns and destruction, four million slaves helped to decide what time it would be in American history.

As W. E. B. Du Bois later wrote, "Of all that most Americans wanted [from the war], this freeing of slaves was the last." But most slaves knew what they wanted out of the war. They sought a sense of independence; they wanted wages, dignity, family security and cohesion, freedom from a master's or an overseer's whims, and freedom from living as both property and the object of another person's will. They wanted release from white people's expectations, from a white man's lust or a white lady's daily needs. They wanted last names, recognition of their own humanity, and control of their children. They wanted labor and livestock that they truly owned; they wanted to live by a clock they themselves could set. They wanted to be, as Douglass once put it, more than a "creature of the present and past," and craved "a future with hope in it."[2] As elusive as it was, the freedom most slaves sought was something they comprehended deep in their experience of its denial.

So the freedom that most slaves imagined was both a long and painful process and a moment, the glorious "jubilee" foretold in their readings and hearings of biblical prophecy. They waited and chose carefully when and how to act while sustaining their expectations of freedom against great odds.

The exact time and location of deliverance, though, had never really been the point. Slaves managed fear with faith; antebellum African American religion was one way people sustained hope of liberation in the face of whatever soul-killing events happened in their own lives. In 1859 a fugitive slave known only as N.L.S., who had just escaped through the efforts of Underground Railroad tactician William Still of Philadelphia, wrote with wonderful syntax to his liberator: "What time was it, when israel went to Jerico i am very anxious to hear for thare is a mighty host will pass over and you and i my brother will sing hally luja i shall notify you when the great catastrophe shall take place."[3] As Union armies and navies penetrated the Southern landscape and harbors, most American slaves seemed to know what time it was. They were willing to be liberated by marching soldiers and intrepid sailors, but they were also ready to free themselves.

John Washington and Wallace Turnage knew what time it was. Both embraced moments of opportunity with youthful courage and risked much to strike out for a new life beyond slavery. The war and the presence of Union armies and navies opened pathways to freedom for them, as it did for so many slaves. But nothing inevitable determined their emancipation. John Washington and Wallace Turnage seized their moments, moving fearfully yet joyfully across rivers and ocean waters out of the Babylon of the Confederacy.

Each man wrote a redemptive story of his freedom. Both men also left us enough of a record to trace the backgrounds of their two quite different, yet compelling and representative journeys. In their stories we can see a revealing convergence of events with personal will, of fate with human choices, that produced heretofore unknown dramas.

Both Turnage and Washington considered themselves quite ordinary; they sought no fame for their courageous escapes. Washington became a house painter and Turnage a nightwatchman, among other occupations. Each would live well into the twentieth century and forge a working-class family that struggled against the barriers of class and racial discrimination in teeming American cities. Yet they possessed the will to write, to make their stories of liberation known, to find readers and garner recognition—even if only within their families—as men who had conquered their condition as slaves, remade themselves as free people, and left a mark on time as best they could. They wanted someone to know they were veterans of the great struggle between North and South, quiet heroes of a war within the war to destroy slavery that Americans were beginning to forget.

Turnage and Washington, whose paths probably never crossed, strove in their own ways to speak for all those folk who want the world to recognize their experience and listen to their memory. Their narratives make an interesting comparison to the extraordinary collection edited and published by Hamilton Holt in 1906, *The Life Stories of Undistinguished Americans, As Told by Themselves.* Holt was an editor at the magazine *The Independent*, which for several years had invited ordinary Americans, mostly immigrants, to tell their stories in brief "lifelets," twenty of which are collected in the book. Immigrant narratives had become a popular genre by 1900, and one purpose of Holt's collection was to demonstrate America's modern, polyethnic, urban society from the "bottom up."[4]

Two blacks appear in Holt's book; like the other storytellers they are types and their names go unidentified. The story of the "Negro peon" reads somewhat like Turnage's woeful tale of enslavement, flight, and punishment. Held for three years "in bondage on the ground" because of perpetual debts in a remote "Georgia peon camp," this approximately forty-year-old man lived in nothing short of reenslavement under horribly oppressive conditions. "I could tell more," he remarks in words reminiscent of Turnage's caution about lurid detail, "but I've said enough to make anybody's heart sick." Similarly, the "Southern Colored Woman," although deeply wounded by white racism and Jim Crow discrimination, might have been describing Washington when she discusses her father, who was the son of his white master. "My father," she remembers, "said after the war his ambition was first to educate himself and family, then to own a white house with green blinds, as much like his father's as possible, and to support his family by his own efforts . . . He succeeded."[5] Even more thoroughly and directly than these turn-of-the-century vignettes, Washington's and Turnage's stories, untouched by editors, provide windows through which we can see how the lives of common people both shape and are transformed by epic events.

Turnage and Washington also show us that the human memory is an inexhaustible treasury of experience and impression ready to erupt into story if there is a way to tell it. But memory can also be a dangerous place to go, as it surely was for both Turnage and Washington. Recollecting slavery was always a risk in a society that so valued progress and success. As Richard Wright wrote in 1941, the slave past for African Americans was potentially a shadow of "shameful identification," a psychological minefield of "self-disgust at our bare lot." In the post-emancipation years even Douglass admitted that forgetting the most traumatic parts of the past might be at times "nature's plan of relief." But he challenged his generation of former slaves to never forget their experience, to face it, and to demand that the nation remember it. "It is not well to forget the past," Douglass warned in an 1884 speech. "The past is . . . the mirror in which we may discern the dim outlines of the future and by which we may make them more symmetrical."[6]

To Douglass, blacks were morally bound to uncover and tell their history, to reshape the national memory by pushing their experience to the center of the story. We do not know precisely why Turnage and Washington wrote their narratives, although I do offer speculations in this book. Their aims were more modest and private than Douglass's. But surely they felt deeply in their souls that their own children, likely the principal audience for whom they wrote, could shape some kind of meaningful future from their mirrors of the past.

The timeworn suspicion that the African and African American past cannot be known because blacks did not leave enough of a written record (a widespread assumption as late as the 1960s and 1970s) has now been put to rest in part by the body of testimony we know as slave narratives. Indeed, for generations after the Civil

War, slavery itself remained a subject steeped in mystery, sentimentalism, and denial, as most whites assumed that a black "silence" on the subject signaled not fear or ambivalence, but a placid acquiescence to enslavement and segregation. The ingrained mythologies from which a society draws its identity die hard; the image of faithful slaves and noble, caring masters endlessly nourished the system of white supremacy that ruled race relations and popular culture in America for three quarters of a century after Reconstruction. The rediscovery of slave narratives in the Civil Rights era and the growing use of the more than three thousand oral testimonies of former slaves recorded by the WPA in the 1930s broke the silence and mapped the way to a new history.

Moreover, interest in slave narratives or the writings of exslaves and free blacks has spiked considerably in recent years with the discoveries of works of fiction by three black women. First came the rediscovery in the 1980s of *Our Nig or, Sketches from the Life of a Free Black,* by Harriet E. Wilson, first published by the New Hampshire native in 1859 and brought back into print with fanfare in the 1990s as the first novel by a black woman. Then Henry Louis Gates, Jr., bought at auction the remarkable manuscript *The Bondwoman's Narrative, By Hannah Crafts, A Fugitive Slave Recently Escaped from North Carolina.* Crafts's autobiographical novel was written, it is believed, between 1853 and 1860. Gates republished it in 2002, creating a great stir among literary critics, historians of slavery, and the general reading public. Finally, in the fall of 2006, William L. Andrews and Mitch Kachun edited and published *The Curse of Caste; Or the Slave Bride: A Rediscovered African American Novel* by Julia C. Collins. Collins, an African American woman who lived in the town of Williamsport, Pennsylvania, had first published an unfinished, serialized version of this work in 1865 in the *Christian Recorder,* the national newspaper of the African Methodist Episcopal Church.[7]

Each of these autobiographical novels has stimulated a rich debate about which is the "first" such work of fiction by a black woman, as well as considerable discussion about how black women of the period represented their emotional and temporal lives, how they gained literacy, how they chose what they read, and how their writing provides access to the inner worlds of slavery and the free black experience. All three of these works were fiction rooted in the personal experiences of the authors, but at least two of the three (Crafts excepted) were mediated to some degree by editors or a publication process. Each was a stunning historical find and offers firsthand accounts of the slave experience. Crafts's story in particular provides, as Gates writes, "access to the mind of the slave in an unmediated fashion."[8] This is precisely the lasting value and historical significance of Washington and Turnage's narratives. They emerge from obscurity almost pure in their unadorned, untouched quality. In this case we have two unedited and authentic ex-slave voices, remembering and recording the events that forged their liberation. In so doing, however self-consciously, Washington and Turnage produced two rare visions of the meaning of emancipation.

The genre of slave narratives is generally divided into three types: biographies, fiction, and autobiographies, with the third category (which includes oral histories) by far the largest. Autobiographies by former slaves were first published in the late eighteenth and early nineteenth centuries and grew in scale as new texts were promoted and printed by the early abolition movement. From the 1740s to 1865, approximately sixty-five autobiographical slave narratives were published in book or pamphlet form. Among them are a handful of classics published between 1840 and 1860 by Frederick Douglass, Harriet Jacobs, William Wells Brown, Solomon Northup, Josiah Henson, Henry Bibb, and a few others. Between the Civil War and the 1920s, approximately fifty to fifty-five ex-slaves published their autobiographies. By far the most influential of the postbellum narratives is Booker T. Washington's *Up From Slavery* (1901), the widely read story of the founder of the Tuskegee Institute in Alabama and the most famous black leader of his era.[9]

Antebellum slave narratives tended to conform to certain structures and conventions. Given the depth of racism in the era, rooted in assumptions of black illiteracy and deviance, pre-1860 ex-slave autobiographers had to demonstrate their humanity and veracity. They had to prove their identity and their reliability as first-person witnesses among a people so often defined outside of the human family of letters. Hence, virtually all pre-emancipation slave narratives include phrases such as "written by himself" or "herself" on title pages, as well as numerous testimonials, prefaces, and letters of endorsement by white abolitionists and supporters. The narratives almost invariably contain a first sentence beginning "I was born," identifying a specific place but often no date of birth. Most also include a photo or engraved portrait of the author up front and appendices containing documents— bills of sale, free papers, newspaper clippings, sermons, speeches, poems—that argue the case against slavery. Most narratives were cast as contests between good and evil, moving through countless examples of cruelty toward slaves and ending in a story of escape. Many are essentially spiritual autobiographies, journeys from sinfulness and ignorance to righteousness and knowledge. On one level, antebellum slave narratives were effective abolitionist propaganda, condemnations of slavery in story form. The very best of them, Douglass's and Jacobs's for example, are also serious works of literature in which the authors psychologically reveal a sense of self and use the truth-telling of art to creatively expose the larger national problem with slavery.[10]

Post-emancipation slave narratives, however, changed in content and form. They still tend to be spiritual autobiographies, often by former bondsmen turned clergymen, and they were written in the mode of "from slave cabin to the pulpit." But postslavery narratives are more practical and less romantic, more about a rise to success for the individual and progress for the race as a whole. Some sport such titles as "From Slavery to Affluence." It is not so much the memory of slavery that matters in the bulk of the postwar genre, but how slavery was overcome by a narrator who competed and won his place in an ever-evolving and more hopeful present. Slavery is now a useable past in the age of Progress and Capital. Frederick Douglass had portrayed slavery in 1845 as a "prison" or "tomb" out of which he fought and willed his own "resurrection." But in Booker Washington's bestselling 1901 variation on the Horatio Alger myth, slavery is a "school" in which blacks have been prepared to compete in the strife of life. Antebellum narratives are saturated with the oppressive nature of slavery and a world shadowed by the past. Postbellum narratives reflect backward only enough to cast off the past, exalt the present, and forge a future. Literacy and reading were precious discoveries in the prewar narratives; after emancipation they were a means to ownership and action. A significant number of post-emancipation narratives were authored by former Union soldiers eager to record their military service and tell their own tale of self-reliant success. Overall, the aim of postbellum narratives was no longer to catalog the horrors of slavery, but to use memoir as a marker of racial uplift and respectability in the age of Jim Crow.[11]

These shifts in the character and purpose of slave narratives make the emancipation stories of Washington and Turnage all the more unusual. They share elements of the classic antebellum slave narrative, but are also quite distinct within the postbellum collection into which they fit. Neither Washington nor Turnage were accomplished writers; they did not have editors to help shape their prose or correct their spelling and grammar. Undoubtedly they were aware of Douglass and others, but these two narratives are untouched creations, boasting neither sponsors nor accompanying documents to buttress their authenticity. Washington conforms to the "I was born" convention in his opening sentence, and his work also carries the author's own title, "Memorys of the Past." Turnage's narrative is untitled, although his daughter seems to have suggested "Adventures and Persecutions." In this publication I have adopted "Journal of Wallace Turnage" as the simplest title in the spirit of the author's own writing. Turnage's opening sentences reflect his reticence as an unschooled writer: "Wallace Turnage's apology for his book. My book is a sketch of my life or adventures and persecutions which I went through from 1860 to 1865."[12] Turnage was largely free of any self-conscious conventions or expectation of style, but his shy apology did not prevent him from telling the dramatic tale of his five escapes on the road to freedom.

Autobiographies are creative acts of recollection, as well as historical documents. Most slave narratives are a combination of memory and imagination on the one hand and descriptive narration on the other.[13] Both Washington and Turnage strove to recreate the one piece of their experience they deemed worthy of telling. Both had to imagine their way into a past of facts— people, events, rivers, roads, and scenes. Each man probably used maps and other sources to prompt his memory and craft his story as truthfully as possible. It is clear that both had a sense of plot— the individual hero struggling against great odds and in the end, with the help of God, the Union forces, and a few friends, triumphing over evil. It is also clear that both possessed a deep religious faith in which God held the fate of the lonely hero in his hands. And each wrote in a present that fueled his desire to remember and record his story.

If either author had secured an editor-publisher to prepare his narrative for the literary marketplace of the 1870s or 1890s, we would be reading quite different texts here. As they are, these narratives provide unfiltered access to the process and the moment of emancipation. In their own personal ways, Washington and Turnage are saying: Here is who I am; here is how I achieved freedom; and here is what it means to me.

Both Washington and Turnage offer apologies for their "ungrammatical" writing. Turnage flirts briefly with a Bookerite notion of his "hard time" making "a man of me," but he very quickly places himself as a purchased commodity in a slave pen in Richmond. Washington hurries right into his earliest idyllic childhood memories of running in "sweet scented cloverfields after the butterflies," adventures that serve as metaphors for freedom before "Slavery" dominates his remembered life. Washington and Turnage wanted to convert suffering into inspiration—for their children and anyone else who might one day read them. They were a little like the boy in Jerzy Kosinski's *The Painted Bird*, orphaned and overwhelmed by extreme brutality and neglect as he travels through eastern Europe during World War II and the Holocaust. Rendered mute by his trauma, he tries to survive by his wits and will. His silence sets him apart in a world of chaos. "His emotions, memory, and senses divided him from others," Kosinski narrates, "as effectively as thick reeds screen the mainstream from the muddy bank." But in the end, through the horror and shattered memory, the boy says, "I opened my mouth and strained. Sounds crawled up my throat. Tense and concentrated I started to arrange them into syllables and words."[14]

In similar ways, Washington's and Turnage's narratives crawled from their memories, as both authors struggled to bring order to their traumatic pasts. It is very hard to know how many other published slave narratives or other forms of literature either man had read before writing his own. But their texts arrive after more than a century of hiding as artful autobiographies crafted in alternately eloquent and broken prose, and as gifts of documentary insight into how slavery was transformed into freedom. In writing their stories, Washington and Turnage tell us more than even they could imagine.

CHAPTER ONE

The Rappahannock River

Day after day the slaves came into camps and everywhere
the "Stars and Stripes" waved they seemed to know
freedom had dawned to the slave.
  —JOHN WASHINGTON, 1873, REMEMBERING AUGUST 1862

John M. Washington was born a slave on May 20, 1838, in Fredericksburg, Virginia. Washington begins his narrative with the wry comment that he "never had the pleasure of knowing" his mother's owner, Thomas R. Ware, Sr., who died before John was born. And he supposes "It might have been a doubtful pleasure." So far as can be determined, Washington also never knew his father, though we can assume he was white. As an autobiographer reconstructing his own youthful identity, Washington says revealingly: "I see myself a small light haired boy (very often passing easily for a white boy)."[1]

With these words Washington recollects the complicated story of so many American slaves—mixed racial heritage. The offspring of sexual unions between black women and their white male owners or pursuers suffered a legacy of confusion, shame, and abuse, but they also occasionally benefited from economic and social advantages, especially in towns and cities. Washington was one of more than 400,000 out of four million American slaves by 1860 who were officially categorized as "mulatto" or other terminology to distinguish a person of some white parentage. From 1830 to the Civil War, the state of Virginia especially had gone to great effort, although unsuccessfully in practical terms, to legally establish a color line marking who was white and who was not.[2] White friends, and perhaps relatives, aided John's education and opportunities early in his life. But in Fredericksburg and elsewhere, due to his mother's status and color, he was considered a chattel slave until the war came.

Exactly who Washington's father was, and how John got his middle initial and last name, have been impossible to trace. A John M. Washington, a distant cousin of President George Washington, lived in Fredericksburg, went to West Point in the 1810s, became an artillery officer, and died in a shipwreck in 1853. But no evidence exists for his patrimony of John. Ware had four sons by 1838, ages twenty-six, twenty-four, twenty, and eighteen. Any of them could have been Washington's father, although only the two younger ones, John and William, seem to have been residents of Fredericksburg at the time.[3]

Washington's story is much clearer on his mother's side. Women determined, protected, and supported John's life chances. His maternal grandmother was a slave named Molly who was born in the late 1790s and owned by Thomas Ware. Molly, called "my Negro woman," is acknowledged for her "faithful service" in Ware's 1820 will, in which he bequeathed her and her children (valued at $600) to his wife, Catherine (who would eventually be John's owner). By 1825 Ware's estate inventory lists Molly and four children; John's mother, Sarah, was the oldest at age eight. Molly would have another four children by the 1830s. In June of 1829 this strong-willed mother misbehaved (perhaps running away) in such a manner that Catherine Ware arranged with a punishment house to execute a "warrant against Molly and for whipping her by contract $1.34."[4] Perhaps Molly's defiance was sparked because her sister, Alice, had just been sold away for $350.

We can only imagine the sorrow and scars in Molly's psyche, a woman whose life was spent nursing white children as well as her own and serving the extended Ware family. But she would live to join her grandson on their flight to freedom in 1862. She died a free woman near her daughter, grandson, and great grandchildren. Whether she departed as a sad or a joyful matriarch, John Washington does not tell us. His silence about Molly may reflect that he was telling only his own heroic story, which did not allow for his grandmother's saga, but it could also represent a part of his family history he was not prepared to expose.

Sarah Tucker, John's mother, was likely born in January 1817. Who the men fathering all these children were remains a researcher's mystery. Sarah probably also had a white father; she is described in various documents as being "bright mulatto" and short in height.[5] Ware did not own any men who could have been either Sarah's or John's father. When Sarah gave birth to John in 1838, she was a twenty-one-year-old who had somehow learned to read and write, a less unusual accomplishment for urban slaves in small households than for plantation slaves.

In 1832, when Sarah was a teenager, Catherine Ware married Francis Whitaker Taliaferro, a plantation and slave owner with four grown children. The Taliaferros had their own slaves and hired others when they needed extra hands, as was the common practice; in 1836 Mr. Taliaferro advertised for "ten able-bodied men for the remainder of the year," offering twelve dollars per month to their owners. The Taliaferros also hired out their own slaves on occasion, including Sarah. With John in tow, Sarah was hired out in 1840 to a farm thirty-seven miles west of Fredericksburg, owned by Richard L. Brown of Orange County.[6]

Washington yearningly describes his eight years in the countryside in the idyllic opening section of his narrative. His mother must have worked as a house slave because he played "mostly with white children." He spent summers "wading the brooks" and climbing ridges from which he could see the "Blue Ridge Mountains" and a "moss covered wheel . . . throwing the water off in beautiful showers" at a mill on the Rapidan River. Among these pleasant memories is his going to a circus at Orange Court House, where he got lost from his family, and his attending services with his mother at the "Mount Pisgah" Baptist church, a large structure "with gallerys around for colored people to sit in." John loved the "tall pines" that surrounded the church and remembers the "cakes, candy and fruits" sold under the great trees on Sundays. He relished his recollections of "corn shuckings," a "hog killing," and a joyous Christmas celebration. He also remembered his mother teaching him the child's bedtime prayer, "Now I lay me down to sleep, I pray the Lord my soul to keep," and the "Lord's Prayer." And perhaps most important, by the time he was eight, Sarah had taught him the alphabet.[7]

Equipped with literacy, if not with good spelling or grammar, Washington brilliantly uses all of these images of nature as backdrop for his descent into the hell of slavery. He employs natural beauty as a metaphor for freedom and a reminder of the terror of bondage, knowing that the glories of nature can both inspire the soul and mock human sadness. He worries at one point that his "minute events" would not "interest" his reader, and then he quickly moves his story forward.

These early years were both easy and painful for Washington to remember. He likely had no memory, though, of his mother's attempt to run away when he was only three. On February 19, 1841, Thomas R. Ware, Jr., advertised in a Fredericksburg newspaper for a "NEGRO WOMAN SARAH." She is described as "about 20 years of age, a bright Mulatto, and rather under the common size." Clearly she had fled some distance and for some length of time, because the notice offered a twenty-dollar reward if Sarah was captured "more than 20 miles from this place."[8] No evidence survives to indicate how and when Sarah was captured or why she fled. Perhaps she simply took flight from the pressures of daily life for a while. Perhaps she was a young, disgruntled woman "lying out," as the saying went, absconding to the woods or another farm to be with her lover. But she was surely a woman of unusual intelligence and resourcefulness if she managed to escape and remain on her own for a period of time.

A recent study of runaway slaves in the antebellum South found that slaveholders' advertisements often described a slave as "proud, artful, cunning . . . shrewd" or "very smart." Historians Loren Schweninger and John Hope Franklin conclude that the typical runaway exhibited "self-confidence, self-assurance, self-possession . . . self-reliance." It was rare for women to run away, especially those with small children. In the database produced by Schweninger and Franklin, based on extant runaway advertisements in five Southern states, 81 percent of all runaways were male. Of the 195 Virginia runaways from 1838 to 1860, of which Sarah would be one, only seventeen (9 percent) were female.[9]

Sarah likely never told her son the story of her flight, although he eventually might have learned of it from others. That Washington had a mother who herself had been a runaway provided a deep layer of silent inheritance, embedded in his spirit if not in his memory. No doubt, both John's mother and grandmother kept parts of their own physical, emotional, and sexual stories to themselves. Perhaps their experience with white men and with rearing children in the desperately insecure world of slavery left them much like Harriet Jacobs, the author of one of the most important slave narratives. "The secrets of slavery are concealed," wrote Jacobs, "like those of the Inquisition. My master was, to my knowledge, the father of eleven slaves. But did the mothers dare tell who was the father of their children? Did other slaves dare to allude to it, except in whispers among themselves? No indeed!"[10]

By 1848 Sarah had four more children—Louisa, Laura, Georgianna, and Willie, all presumably born on the Brown farm. Sarah, like Molly before her, now had a growing flock of young to worry over and feed. She would have intuitively understood Jacobs's assertion that "the mother of slaves is very watchful. She knows there is no security for her children. After they have entered their teens she lives in daily expectation of trouble." Sarah would eventually achieve freedom with her son, but her more than two decades as a slave and mother made her undoubtedly one of the women whom Jacobs spoke for in her eloquent and terrible questions: "Why does the slave ever love? Why allow the tendrils of the heart to twine around objects which may at any moment be wrenched away by the hands of violence?"[11]

As the Mexican War wound to a conclusion in 1848, a crisis over the expansion of slavery exploded in American politics because of massive land acquisitions in the Southwest. Revolutions against monarchies broke out all over Europe. And the Brown farm was sold. Ten-year-old John Washington moved back to Fredericksburg with his mother and four younger siblings. John lived as the house servant of Mrs. Taliaferro, now a widow, while Sarah lived with the four younger children in a house on George Street near the Rappahannock River. John describes his mother as "sent to live to herself . . . without any help from our owners (except) doctors bills." Mrs. Taliaferro's son, William Ware, was a teller at the Farmer's Bank, where she may have boarded for a time. Washington writes that he "was dressed every morning . . . in a neat white jacket and pants and sent up to the Bank to see what Mistress might want me to do." As her servant boy, John was sometimes forced to "sit on a footstool, in her room for hours . . . when other children of my age would be out at play."[12] He relates such childhood stories in a chapter starkly entitled "Slavery," as though a veil had descended over the innocence of his youth.

But that innocence had begun to vanish while on the Brown farm. From those otherwise bucolic years, Washington recollects his "first great sorrow": He watched a coffle of slaves, "formed into line, with little bundles strapped to their backs . . . marched off to be Sold South away from all that was near and dear to them." In the 1840s, slave coffles were a common sight, as slaves were traded from the upper South to the deep South, where the demand for their labor had exploded. The fear of such a sale always loomed over Washington and his kin. In the coffle, John watched families disintegrating before his eyes. "I shall never forget the weeping that morning," he remarks, "among those that were left behind each one expecting to go next." He could take solace, however, from his beloved mother's instruction in reading. John gratefully acknowledges Sarah's keeping him "at my lessons an hour or two each night."[13] At the age of ten he was as equipped as his struggling mother could make him in the insecure world of slavery. Washington could read; he had learned to wear pants; he was honing his negotiating skills with his mistress and other white people. And he now lived in a city—where the boundaries of slavery were permeable.

"A city slave is almost a freeman," wrote Frederick Douglass in his *Narrative*, "compared with a slave on a plantation." Washington's story confirms Douglass's ironic claim. In Southern cities, the lines between slavery and freedom did, indeed, become blurred. White Southerners complained endlessly of their dilemmas with urban slavery, where the master-bondsman relationship lost so much of its rural plantation routine and certainty. "The problem," writes the historian of urban slavery Richard Wade, "was not what happened in the factory or shop but what happened in the back street, the church, the grocery store, the rented room, and the out-of-the-way house." Cities "corrupted" slaves, wrote a Louisiana planter, attracting them to the "worst habits." But a group of free blacks in Richmond saw it differently; in their view, they had merely "acquired town habits." The Southern journalist J. D. DeBow saw the issue clearly. "The negroes are the most social of all human beings," he said, "and after having hired in town, refuse to live again in the country." A white Northern traveler, John S. C. Abbott, described most insightfully why cities did not bode well for slave obedience: "The atmosphere of the city is too life-giving, and creates thought," he remarked.[14]

This was precisely the experience, both fortunate and anguished, of John Washington. From his early teens on, Washington became very social, and he developed lasting "town habits." Eventually his freedom would be the product, in part, of his own brand of virtuous corruption learned in the interstices of urban slavery.

Some of "what happened" in the back streets and groceries started with the growth of complex connections between slaves and free blacks, a phenomenon observed in every city. The free black population significantly outnumbered the slave population in every antebellum Southern city despite restrictive residency laws. As precarious and marginal as their lives were without political rights and with curbs on their property ownership, free blacks were nevertheless crucial to urban economies, and they developed a good deal of autonomy despite discrimination. Above all they were a social and moral threat to the slave system. Free blacks were a "plague and pest in the community," declared a New Orleans newspaper, "elements of mischief to the slave population." In cities of all sizes in the South, slaves and free blacks shared housing, jobs, churches, social gatherings, friendship, marriage, and blood kinship. That John Washington—like Frederick Douglass and others before him who had escaped from cities— would marry a free black woman and grow up in her social circle was not at all unusual in the world of urban slavery.[15]

The bloodlines and folkways of black and white people were also interwoven in the slave society of Fredericksburg. John's personality, his heredity, perhaps his very worldview were a remarkable example of the interdependence of black and white cultures that Ralph Ellison famously described a century later. "Southern whites cannot walk, talk, sing, conceive of laws or justice," wrote Ellison in 1963, or "think of sex, love, the family, or freedom without responding to the presence of Negroes."[16] With an unknown white father, and both black and white relatives around him, John came of age as a slave, but with his eyes on a way out. In the shops, workplaces, churches, taverns, homes, and streets of a Southern city, John was on the surface both black and white, both slave and free. In time he probably possessed a higher degree of freedom of movement about town than he admits in the narrative; he entered stores, did financial transactions for his mistress, and talked among friends—black and white. With his responsibilities as a servant and laborer came everyday reminders of the possibilities of life beyond slavery. Washington was cornered. Legally he was someone's property, a creature of others' pride, profit, and will. But he was a young person of talent, an extremely valuable and agile spider in slavery's tangled web; and he spun his share of that web as it was in turn spun around him. He knew that his owner's dependence on him provided his best hope for independence. He learned to work, wait, and, above all, to deceive his mistress.

During these years, Washington's owners valued him highly for his intelligence and skills; their problem was how to control him. As Mrs. Taliaferro tried to restrict his movement and his will, Washington became a "confirmed Liar," a declaration proudly made by the authors of many slave narratives. Mrs. Taliaferro was a "strict member" of a Baptist church attended by both blacks and whites. Slaves sat in galleries on each side, "free colored" sat in the center gallery, and whites sat on the main floor. "I used to have to sit where the old lady could see me," wrote Washington, "as proof that I was there." As proof that he had dutifully attended the Sunday afternoon gatherings held by blacks in the church basement, his mistress would demand that he bring home knowledge of the text used by the preacher. Resorting to "all kinds of sub-

tifuge," John would hang around the church door, memorize the
text as the minister announced it, then escape to the riverfront
with the other boys to play marbles, swim, or row any boat they
could find. These boats had a lasting impact on his imagination
and his life. "I had the greatest love for the river and boats and
such risks," he recalls with bravado in the narrative. Upon returning home he would name the biblical text for his mistress to garner her approval.

Washington became a "thorough hater" of Mrs. Taliaferro's
church; in time, he would come to prefer and even love black-
church culture and practice. By 1855 a new whites-only Baptist church had been built in Fredericksburg. The old structure
quickly took on the name "African Baptist" and later became, as
it still is today, the Shiloh Baptist Church. In 1856, when he had
turned eighteen, John was baptized in the Rappahannock River
and became a member of Shiloh Baptist, along with at least two
of his siblings.[17]

But Washington faced many "trials of slavery" as he struggled
through adolescence. He was "left alone," as he titles chapter three
of the narrative, around Christmas 1850, when his mother and
four siblings were hired out to Staunton in western Virginia, several hours from Fredericksburg by train.[18] Mrs. Taliaferro, probably for financial reasons, sent Sarah and the four younger children
to live with and work for a Reverend Richard H. Phillips, who was
married to her niece, Eleanor Thom. Mrs. Taliaferro may have
owed some sort of favor to her niece and her husband, but like so
many slave owners, she needed money more than Sarah's labor.
Reverend Phillips was an Episcopal clergyman and principal of a
girls' boarding school, the Virginia Female Institute (now Stuart

Hall). Sarah worked as a seamstress or laundress at the school, and there is no record that she ever returned to live in Fredericksburg before John's escape in 1862.

The night before this forced separation, Sarah went to John's bed and lay down next to him. As their "tears mingled," his mother begged him to "remember all she had tried to teach me, and always think of her." Washington describes this moment in his life as one of profound loss and loneliness. "Whose hand (patting) me upon my head," he lamented, "would sooth my early trials?" He made this separation from his mother a pivotal point in his story. "Then and there," he declared with a memoirist's retrospective certitude, "my hatred was kindled secretly against my oppressors, and I promised myself if ever I got an opportunity I would run away."[19]

Such a callous severing of a mother's connection to her own child provides a staple theme in antislavery literature, especially in slave narratives. "My mother and I were separated when I was but an infant," Frederick Douglass wrote, "before I knew her as my mother." Douglass invented images of his mother in his autobiographies, likening her to an Egyptian queen he saw in a book. Charles Ball, like Douglass, spent his childhood years in slavery on Maryland's eastern shore. He and his siblings were sold away from the clutches of their mother's arms on the same day that she was sold south to Georgia. "The horrors of that day sank deeply into my heart," remembers Ball. His mother's buyer had to whip her about the shoulders to force her to release her five-year-old son. As the cruel master "dragged her back toward the place of sale," Ball remembers his mother's screams as well as her tenderhearted soul. And Josiah Henson, another Marylander, recounts a wrenching scene from his childhood: The death of their master meant the selling off of his estate, and his mother, "paralysed by grief," falls to her knees and clings to a buyer's ankles, begging that her youngest of six children (Josiah) not be sold. Henson described this childhood hell, watching as his mother "crawled away from the brutal man."[20]

The twelve-year-old Washington was luckier than many others; he would see his mother once more before adulthood. And he was able to correspond with her, given their mutual literacy. This separation, and the fact that whites on either end might write and read the letters John and his mother exchanged, forced him to work on his reading and writing. John read *Harpers Monthly*, copies of which he found in William Ware's room when he cleaned; the short stories were particularly enticing. Some of Ware's white friends also helped John with his spelling, and his uncle George (his mother's brother) assisted him with his penmanship. Finally, with the purchase of a twelve-cent copybook, he practiced his alphabet. These letters already set John apart from other slaves; he was becoming a creature of written language—something less than one in ten slaves achieved—in order to still know his mother, as well as to imagine his way beyond a slave's life.[21] The scars left by his long separation from his only parent and all of his siblings had to have influenced John's later determination to gather and protect his own family.

Washington got to see his family once more in 1852, two years after they left. He had contracted a case of poison oak after a rollick and a swim in the Rappahannock with some other boys in stolen rowboats. He became so sick that his mistress decided to send him out to Staunton to recover with his mother. The trip by railroad and stagecoach over mountains and into the Shenandoah Valley thrilled him. The three months with his family were joyous, but ended abruptly in autumn as Reverend Phillips informed him he had to go back to Fredericksburg. In the story of his journey home, Washington combines a deep fascination for railroads and adventure with a sense of humor, mentioning the white folks' worries that he might make an "escape and get to a free state . . . provided I would do such a disgraceful act." He also turned this episode into an abolitionist statement about the "white man's power and oppression of the Colored slave" and a sarcastic reminder that Phillips had to find "some white person" to serve as guardian on the trip to "take charge of my body." He did not miss the chance to link racism to slavery. Amid rich detail about train accidents, engine names, speeds, and railroad junctions, Washington sounds like a fourteen-year-old, bursting with independence and a traveler's excited observations of all that was new. He rode in "the Niggers Car," and was greeted in Fredericksburg by white people "relieved from anxiety of my possible escape to freedom."[22] Washington sometimes struggled with grammar, but irony flows from his memory like the Rappahannock itself.

Washington's narrative is very much a coming-of-age story, a remembrance of what it meant to be a slave, but it is also about a teenager's falling in love and converting to a personal religious faith—a saga of the human impulse toward freedom bursting from what he calls his "close imprisonment." Mrs. Taliaferro constricted Washington's movement, rarely permitting him to "pass the limits of the lot." He is a trapped teenager eager to join "rollicking fun loving companions," but he is constrained by his owner's constant fear of his escaping. Just as Douglass turned the sailing ships he witnessed on Chesapeake Bay during the depths of his teenage enslavement into "freedom's swift-winged angels" beckoning to him as symbols of an unreachable liberation, so

Washington converted the "beautiful surrounding country" around Fredericksburg into his natural metaphor for unattainable liberty. "Imagine a boy," he says, allowed only to gaze out at nature's wonders from his perch in an open window: "The sweet scent of clover locust blossoms, huneysuckle, apple, cherry, and various fruit trees almost ripened, and all nature clothed in beauty that cannot be describe[d]." Slavery, he argues, not only constricted the body and mind, but even the senses. He was "not permitted," he complained, "to go out and see and smell the work of Him, who created all things."[23]

Washington returned from Staunton to a life sometimes even more constricted than before. As he got older Mrs. Taliaferro tightened his bonds, her fear of his escape leaving him little freedom of movement when he was in town. However, he was allowed to attend church fairs and revivals conducted at the African Baptist church, just two blocks downhill from where he lived in an upstairs room at the Farmer's Bank. At one of these revivals—he remembers the date as May 25, 1855—he experienced a personal conversion, a deep sense of "sin," and worry over the "salvation of [his] soul." Washington's Christian faith was extremely important to his sense of identity, his self-confidence, and his quest to envision a future beyond slavery. He likened it to "burning coals, fanned by the breeze." He declared himself henceforward a close reader of the Bible, and he was baptized in the Rappahannock on June 13, 1856. When Langston Hughes later wrote his poem "The Negro Speaks of Rivers," John Washington was surely one of those former slaves about whom he mused and for whom he spoke: "I've known rivers/ My soul has grown deep like the rivers."[24]

At one of the church fairs Washington met the love of his life, Annie E. Gordon, a black girl, born free in 1842. Annie, too, was light-skinned and was listed as "mulatto" in the 1850 census. She was shy and four years younger than John, who first courted her with a Valentine message passed on through a friend. They saw each other on the streets of Fredericksburg and at as many church picnics and "partys" as possible during the 1850s. Washington details some of their courtship in a remarkable document, half essay and half diary. He tells of awkward first meetings with this "beatyful girl" when they were eighteen and fourteen, and how his friend, Austin, played the role of intermediary in the courtship. The dreamy young "lad," as he calls himself, was "troubled" by what seemed like Annie's indifference to him. But they began exchanging notes, and soon this grew to a "very pleasant correspondence." Literacy had more than one value to a slave in Virginia; it helped a young man in love. "I kissed the note and blessed the hand that I supposed had written it," he confesses.[25] As in Victorian fiction, he nearly loses Annie to another young man to whom he hears she is engaged. John, too, gave his affections to another girl for a time.

But at the Christmas fair of 1856 he spied Annie across the room, and her "smiles and bewitching manners" floored him. The rest of the evening, John "seated or knelt by Annie or promenaded with her till it was time to leave." In what seems to be a piece of diary manuscript from 1858, he again records his cravings for Annie's attention and love. Not even the "cherry red lips in the most tempting little pouts within an inch of my own lips" of a girl named Sallie could distract Washington from his pursuit of Annie. He describes petty lovers' quarrels: He "got mad," and she "laughed and was crying at the same time," as they pushed and grabbed each other, learning how to pet and touch. The diary entries end on July 16, 1858, with John's disappointment over a "walk down 'Lovers Lane'" that is apparently cancelled by rain.[26]

Annie eventually gave in. On January 3, 1862, nine years after their first meeting, John and Annie were married at the African Baptist church by a white minister, the Reverend George Rowe, himself a slave owner. Recognized marriages among blacks, slave and free, were not uncommon in cities. In some cities, such unions were acknowledged by city ordinances, and mayors conducted marriages in others. But the instability and pressure of slavery placed all black marriages in the South in jeopardy, and marriages between slaves and free blacks were particularly vulnerable. John and Annie were lucky to the extent that he escaped months after their wedding; they did not have to live very long negotiating the boundaries of slave and free status or face for any extended period the fear of John's sale.[27] This marriage across the line of slavery and freedom in the midst of war demonstrates the complexity of that peculiar institution teetering on the brink of destruction. And Washington's diary fragments offer a rare view into courtship between young blacks in the mid-nineteenth century as they pursued the oldest of human urges in a slave society.

Washington's literacy, his connections with free blacks, and his urban environment all contributed to his escape. But another critical circumstance emerged when he was "hired out" several times between 1859 and 1862. As an urban slave of an owner who needed extra money, John could accumulate both skills and cash as a hired laborer. Among the 246,981 enslaved Virginians in 1860, he was among the approximately 25,000 who were hired out. In 1859, Washington went to work for the William T. Hart family, who lived only four doors away from the small cottage (which still

stands) where he boarded with Mrs. Taliaferro at the corner of
Hanover and Prince Edward streets. He spent the year driving
horses, tending a garden and the family's one cow, and perform-
ing all manner of odd jobs. Washington remembers his hiring out
cheerfully since all restrictions on his time and movement ended,
and his opportunities to "make money . . . increased tenfold."
Hidden in the heart of the slaveholders' defending their system as
"paternalistic" and alleging a familial relationship based on the
masters' obligations of care and provision was an often silenced
but rampant pecuniary desire. Fanny Kemble, a Southern woman
who grew to hate the slavery she was once forced to embrace by
marriage, remarked that this common practice of slave hiring only
verified the average slaveholder's primary goal "to get as much out
of them [slaves], and expend as little on them, as possible."[28] Although his owner was caring and kind, especially about his health,
Washington quickly learned how to keep score in this brutal game
of give and take.

On January 1, 1860, the twenty-one-year-old Washington
was hired out again, this time to the Alexander and Gibbs tobacco factory in downtown Fredericksburg across from the train
depot, only a few blocks from where he lived. Tobacco factories
in Virginia towns and cities commonly employed "hired" slaves.
For a year, John worked on "task" as a "twister" of sixty-five to
one hundred pounds of tobacco per day. Each worker was given
a designated goal for his labor (his owner was paid for his time);
beyond that expectation the slaves were paid extra money for any
higher production, allowing the worker in some weeks to garner
four to five dollars for himself, a significant sum in 1860 dollars.
Washington's 7 A.M.-to-6 P.M. work at the factory felt "more like
freedom" than any time since his early childhood. In spite of the

constant "din" of the factory machinery and the overseers whipping slaves who did not complete their tasks, Washington gained
a sense of adult autonomy through work and completed, measur-
able labor.[29]

In the tobacco factory, as at church, Washington gained a
sense of black solidarity and cultural identity. He remarks that on
any day one could hear above the noise "some of the finest singing
known to the colored race." Washington did not record any lyrics,
but his experience surely fits a prominent folklorist's conclusion
that "by incorporating the work with their song, by . . . co-opting
something they are forced to do anyway, they make it *theirs* in a
way it otherwise is not."[30] Washington's paths to freedom were
many—psychological, geographical, and spiritual. But in one important sense, he found ways to *own* his labor and *work* his way
to liberty.

Washington was hired out a third time, again on a January
first. According to Washington, the secession of South Carolina
in December 1860 and the threatened end of business between
North and South forced the closing of the tobacco factory. In this
case he may have sought the arrangement as much as his owner
forced it. He had long wanted to see Richmond, fifty miles to the
south, and other blacks had told him it was a "good place to make
money." In the custody of a hiring agent, Washington went by
train to the biggest city he had seen and was promptly hired out
to a tavern owner, a Greek immigrant named Speredone Zetelle.
After six months Zetelle sold the "eating saloon" to a German
immigrant, Caspar Wendlinger, for whom Washington worked
for the rest of 1861 in the city that became the capital of the
Confederacy with the outbreak of the Civil War. Working only
blocks down the hill from the capitol square—with its magnificent

equestrian monument to George Washington—and the Virginia capitol building, designed by Thomas Jefferson, John watched as "thousands of troops . . . from all parts of the South" marched through and bivouacked in the city. He and his fellow black laborers thought it an "impossibility" that such an army "could ever be conquered." Washington read the newspapers with great zeal, marking news of great battles like first Bull Run, in July 1861, and of tidings about slaves escaping to Union lines or to the "free states."[31] Possessing literacy, a bit of autonomy, and a few dollars in his pocket, Washington was inspired by the secession crisis and impending war to imagine new links between dreams and reality.

It is not likely that the fourteen-year-old Wallace Turnage and the twenty-three-year-old Washington crossed paths in the streets of Richmond, where Turnage was sold to a slave trader and jail owner, Hector Davis, in 1860. Washington's tavern owners, Zetelle and Wendlinger, sent recalcitrant bondsmen to a nearby slave jail for whippings at the price of fifty cents per slave, and that jail was likely Davis's at Franklin and Fifteenth streets. Some combination of Washington's personality, skills, and perhaps even his light skin kept him from such a bloody fate. But the many protestations by officials of the Confederacy in the spring and summer of 1861 that they were not going to war for slavery, but only for their right of "self-government" and constitutional "liberty," would have been a hard sell down in the streets along the James River, at the slave auction house, or in the tavern kitchen where Washington logged his long days and nights.[32]

While in Richmond, Washington exchanged numerous love letters with his soon-to-be wife, Annie, back in Fredericksburg. Only one of the letters survives, but it is an extraordinary record of young love, fraught with ornate language, anxiety, plans for marriage, and promises of loyalty. "Oh if I was home with you to night how happy I should be," John writes. "I would endeavor to cheer thee." John had attended a party where other girls were guests. But he promises that it "will be my last certainly for I do not like partys where you are not." He vows his devotion. "Between you and me there is and must ever be sincerity and truth. You tell me of unhappy hours for my sake. I beseech you to be easy and assured that the memory of thee ever admonishes me to be careful." In this affected language John declares his love and his anxiety "to marry now more than ever." He says his mind is "better settled," and he "really wants someone to call wife." Devoid of a father, cut off for years from his mother, and his siblings living at a great distance, Washington surely was desperate to find the love and human connection that young Annie might provide. A little security of the heart in his utterly insecure world of enslavement and impending war was a prospect that made him gush with excitement: "Oh my own sweet Annie I know you love me and I am so proud of it, and it sends a thrill of joy to my heart."[33] Soon John would be back in Fredericksburg planning a wedding, as well as one other form of joy.

With war mobilization swirling all around, Washington was given a pass to go back to Fredericksburg for Christmas in 1861 with the understanding that he would return to the tavern on January 1. But he never came back to the Confederate capital. In Fredericksburg he found a new "hiring"—or a "home for the year of 1862"—at a hotel tavern called the Shakespeare House, a block and a half from the Rappahannock and the African Baptist church. To the displeasure of Mrs. Taliaferro, who wanted him to return to Richmond, and his bride's parents, who apparently judged John unworthy of their daughter, he married Annie Gordon at

the African Baptist on January 3, 1862. Working now as a bar-
keeper and steward at the hotel, Washington set his eyes and
hopes northward, as Union troops crept closer to his hometown,
which rested approximately halfway between the two adversaries'
capitals. "The war was getting hoter every day," writes Washington, describing how the Yankees moved slowly southward in February and March of 1862.[34]

As the war bore down on Fredericksburg, some whites began
sending their slaves to the countryside, while others, as Washington reports, "were sent into the Rebel Army as Drivers, Cooks,
Hostlers, and anything else they could do." Some neo-Confederate
enthusiasts are currently fond of claiming that thousands of slaves
served as loyal "black Confederates" in the South's cause. Washington's eyewitness testimony offers yet another refutation to this
effort to legitimize the Confederacy in modern memory.[35] Moreover, Washington's narrative about the crucial events that led to
his emancipation carefully demarcates the lines of loyalty between
the Union and Confederate causes in his immediate experience of
the war. He conducted a veritable psychological war with his
owner and his employers over his own fate, as well as that of *their*
Confederacy.

The proprietors of the Shakespeare, George Peyton and
James Mazeen, had joined the Thirtieth Virginia Infantry as officers. In April they told Washington that the hotel would close
because of the approach of the Yankees and that John would have
to accompany Peyton to Salisbury, North Carolina, as his servant.
Admitting that the proprietors of the hotel treated him well in his
quest to make his own money, Washington nevertheless "resolved
not to go," while he "made them believe" he would remain loyal.
He feigned fear of the Union troops when whites were listening

and told Mrs. Taliaferro that he would join her the next morning in her journey as she frantically packed her "silvery spoons" for flight to the countryside. Wild stories abounded in Fredericksburg, with masters telling their slaves that the Yankees would "send me to Cuba or cut off my hands." John ignored all such warnings and never intended to make good on his "yes Madam" to his mistress.[36]

In this state of excitement and chaos, Washington's long-sought opportunity arrived on April 18. As the Union army reached Falmouth, about a mile upriver, the jack-of-all-trades at the Shakespeare Hotel could hardly contain his anticipation. Only ten days after the shocking news of the bloody battle of Shiloh (fought in southwestern Tennessee) spread across the nation, federal forces moved on Fredericksburg from the north.[37] John began to hear and see the Union soldiers; the loud and destructive beast of war had brought the means of his liberation to the banks of his beloved Rappahannock. He had already voted for his freedom; now he just needed some help.

The city was of "manifest importance," wrote General Irvin McDowell, the commander of the Union forces lurching toward Fredericksburg. In McDowell's April 18 dispatch he reported that his troops had "occupied the suburbs" at Falmouth, but could not prevent the Confederates from burning the two bridges across the Rappahannock as they evacuated the town. As the two armies collided, many slaves began to work in their own interests and not in service to their masters' flight. Some were valuable informants to Union officers. Brigadier General Christopher C. Augur, who commanded the brigade that took Falmouth, acknowledged numerous "reports of the negroes" in helping him discern Confederate troop locations, barricades, and plans to burn the bridges.

And Confederate officers both used and feared blacks as carriers of intelligence, with one complaining that "there are so many negroes to inform against me that I shall have to move with the utmost precaution."[38] Armed only with hope, skills, and information, John Washington seized his moment and joined the ranks of the Negro informants.

During breakfast at the Shakespeare Hotel, the dining room was "crowded with boarders" as a cavalryman ran in, shouting, "the Yankees is in Falmouth," touching off "wild confusion" and a hasty retreat of all the whites. Amid "hurried words and hasty footsteps," the Shakespeare's owner gave Washington a "roll of banknotes" with which to pay off all the servants, and ordered him to close up the hotel and take the keys to a safe place. With a delighted sense of irony, Washington reports that all the whites were in the streets of Fredericksburg and heading southward out of town, while all the blacks in and around the hotel were "on the house top looking over the river at the Yankees for their glistening bayonets could easily be seen." Surrounded by turmoil and uncertainty, Washington nurtured inexpressible "new born hopes." Soon the whites were all gone or shut up inside, while the streets of Fredericksburg swelled with black people "in the best of humor."[39]

In this scene on a sunny April day in a Virginia town, the meaning of the Civil War was thrown into full view. As soon as John had put everything in order at the hotel, he gathered all the black "servants" in the bar, paid them their wages, poured a round of drinks for everyone, and lifted a toast to the "Yankees health." He told his fellow "hired" slave workers to "go just where they pleased but be sure the Yankees have no trouble finding them."[40] The sense of humor with which Washington recollects this series of events no doubt thrived on his retrospective repose, but it also contains deep truths about emancipation: Many American slaves fully grasped both the gravity and the irony of their acts. They were ready realists searching for an earthly manifestation of a dream.

As Washington left the Shakespeare Hotel and walked one block down toward the Rappahannock, he noticed a large crowd gathered near the bridge, its "smoking timbers," according to a Union regimental historian, still "dropping piecemeal into the river." A Union captain, J. P. Wood, had crossed in a rowboat to demand the surrender of the city from the mayor and the town council. Washington watched and overheard the exchange as the Union officer offered "unconditional" terms before he went back across the river to his waiting battalion of horsemen. In this excitement, a constable ordered the throng of blacks to go home, but Washington joined his cousin, James Washington, and another free black man, and headed north along the river road toward the music of Union bands that they could hear playing the "Star Spangled Banner" and other patriotic airs. They walked more than a mile until they reached Ficklen's Bridgewater Mill; there they veered downhill toward the river, where they saw numerous Union sentinels on the other side. One of the Union troops shouted out to Washington's little group, asking whether they wanted to cross, to which John shouted back: "Yes I want to come over!" The signal importance of this moment in Washington's life is demonstrated by the stunning hand-drawn map of the city and riverfront of Fredericksburg that he includes in his narrative.[41] In the detail of the map, it is as if Washington is declaring his need to never let this memory recede from his mind—nor from the consciousness of his family.

Within minutes soldiers had escorted Washington and his companions to the north side of the Rappahannock, the most important—if one of the briefest—journeys John had ever taken on that river of his dreams. As they disembarked from the boat, Union soldiers surrounded them, asking all manner of questions about Confederate forces and conditions in Fredericksburg. Washington had "stuffed his pockets with rebel newspapers" and he distributed them all around, much to the delight of his new hosts. Most poignantly, Union soldiers at first assumed that John was a white man and were astonished to learn he was "a colored man . . . and a slave all my life." "Do you want to be free," one of the soldiers asked Washington. "By all means," he replied. One of the Union men informed him that he would "no longer belong to anybody" since Congress had just passed emancipation in the District of Columbia two days earlier, on April 16. "Dumb with joy," Washington thanked God out loud and laughed.[42]

His two companions recrossed the river into Fredericksburg before nightfall, but John remained in the Union camp, befriended by a corporal, Charles Ladd, of Company H, Thirtieth New York Volunteers. That night, his first in freedom, Washington shared meals with the Union soldiers, wandered around their camp, and eventually slept indoors on a wood bench at the house of Eliza Butler, a free black woman in Falmouth who gave over rooms to the troops. His memory was always ready with symbolism and irony. "This was the first night of my freedom," Washington records with double underlining. "It was Good Friday indeed the best Friday I had ever had."[43]

The next morning the freedman witnessed the "solemn and impressive" burial of seven Union soldiers who had died in the battle to take Falmouth. The open graves were dug in the old town cemetery, where "some of the oldest and most wealthy of the early settlers" had been laid to rest. Birthplaces in the British Isles "could be dimly traced" on the tombstones. In such solemnity Washington seems to find his own peculiar poet's voice: "Amidst grand old tombs and vaults serenaded by noble cedars through which the April wind seemed to moan low dirges, there they was now about to deposit the remains of what the rebels was pleased to term the low born 'Yankee' side-by-side." Prayers "wafted to the skies" and hymns were sung, with everyone in tears. Then the former slave remembers the "sound of the earth falling into those new made graves."[44] No young man in Washington's circumstance could ever forget the direct relationship between those Union dead and his new freedom. Witnessing the terrible somberness of his first military burial, Washington stored sensory memories that would compel him one day to write. And in the sound heard by a newly freed slave of the dirt falling on those Northern soldiers' wooden coffins we today can hear echoes of exactly what that war was about.

However, blacks entering Union lines in the late spring of 1862 were still awaiting the enormous military and legal shifts in the character of the war that would bring genuine freedom. They were dependent on the attitudes of Union officers and enlisted men. J. Harrison Mills, the historian of the Twenty-first New York, the unit with which Washington first found refuge, recorded at some length the flow of "numerous contrabands" into the regiment's camp during those April days. He reveled in racist mimicry of the slave dialects that white northern soldiers encountered for the first time: "Lawd! Didn't dem fellus go . . . yah, yah, massa, tought de debbil comin—sure—massa!" But the blacks swarming into their camp were more than merely a source of comic relief. Mills wrote that they were "sharp bargainers" who brought much-needed food and supplies to the hungry soldiers. Referring to all the freedmen as "Cuffee," the regimental chronicler mused that he [Cuffee] "could not clean a gun properly," nor could "ebony substitutes" do "guard duty," but "Cuffee could throw a lively meal." "Poor Cuffee" also entertained the troops with his "weirdly exciting . . . wild, fervid, religious dances."[45]

These "darkey" tales in regimental histories and in soldiers' letters are commonplace, but in this instance, we also have John Washington's own testimony of this war-induced interracial cultural encounter. He did not report whether he participated in any dancing or singing, but he was supremely proud to declare that he would now "be a slave no more," that he could "claim every cent" he earned, and that the men at the soldiers' mess loved his "beef hash." Washington was no one's Cuffee; he was too busy converting his freedom into tangible independence. Informed that he could "get a situation waiting on . . . officers," Washington seized the chance and was hired as a "mess servant" (cook), first for General Augur's staff, and then for the division commander General Rufus King. Within three weeks, Washington was fully in charge of King's kitchen at eighteen dollars per month.[46]

Relishing his new acceptance among the Union troops, Washington virtually became, by his estimation, a member of the army. "I succeeded beyond my expectations," he writes; and he particularly glories in recounting his ride on horseback among high-ranking Union officers as a "guide" when they entered Fredericksburg in early May to arrest "prominent rebels" and establish General King's headquarters in (to Washington's great delight) the Farmer's Bank, his old residence. Here was a former slave, asked by Union officers to "point out each place and to name each person required." How the tables had turned, as Washington proudly helped King's command identify and arrest seven white Confederate sympathizers and escort them off to Washington, D.C.![47]

A revolution was underway in northern and central Virginia, and Washington was only too willing to play his part. He admitted to seeing "some few rebel sympathizers" among blacks in Fredericksburg. But he described "hundreds" seeking passes and transportation of any kind northward, making "their escape to the free states." Four months before Lincoln issued the Preliminary Emancipation Proclamation that freed the slaves in the Confederate states, Washington served as witness to the dissolution of slavery by the action of the bondsmen themselves. "Day after day the slaves came into camps and every where that the 'Star and Stripes' waved they seemed to know freedom had dawned to the slave."[48] Washington uses his narrative to tell the central event of his own life; but he seems keenly aware that a transcendent national history was at stake as well. His retrospective depictions of the emancipation process during that summer of 1862 are some of the most vivid we have from the pen of a freedman. Now and then, he names the truth at the heart of emancipation: In the face of chaos and hardship, slaves seized their liberty whenever and wherever they were given a reasonable chance.

As King's division suddenly moved northwestward in late May to stop Confederate general Stonewall Jackson's extraordinary Shenandoah Valley campaign, Washington left his young wife behind in Union-occupied Fredericksburg and marched with the infantry. Along with several other blacks, he served as a hostler and general camp aide. John now experienced some of the hard marching and privation of common soldiers on the road and through

fields from Catlett's Station to Haymarket, Gainesville, and Warrenton. Camping in abandoned farmhouses or in muddy fields,
Washington used his knapsack for a pillow with "pistols under our
heads and our sabers close by for immediate use in case of an attack" by Confederate "Gurrilas." It is doubtful whether he carried
a pistol and sabre as a camp servant, but he wants his readers to
know that he was "one of the boys." And his mention of the threat
of "Mosby and his gang," referring to John Singleton Mosby, the
famed Partisan Ranger who led his own band of Confederate irregular cavalry and wrecked havoc on Union forces all over northern Virginia, is a trick of memory and an embellishment. Mosby's
Partisan Rangers were not commissioned until January 1863—in
May and June of 1862, Mosby was a guide for J. E. B. Stuart's Confederate cavalry defending Richmond. But this was John adding
his own small leaf to the thousands of soldiers' adventures that
were published in the decades following the war.[49]

In recounting the march, Washington exercised his sharp
sense of irony—noting the contrasts between war and peace, between slave-built wealth and splendor and a Yankee army of con-
querors and liberators. As they reached Warrenton, the former
slave helped set up General King's headquarters in a hotel and observed the beautiful "scenery" around the town, "mountains, hills,
and valleys . . . covered with splendid vegetation this time of year."
The war had brought him near a place about which he must have
heard stories: the "Fauquire White Sulphur Springs," a place
"much frequented before the war by the wealthy in search of
health and enjoyment." The health of the entire society that was
conceived by the slaveholders who visited the springs was now
John's subject. On what was likely Sunday, June 1, 1862, Washington attended a church in Warrenton with a contingent of Union

soldiers. Just after the opening hymn, a prayer, and a scripture reading, an orderly interrupted the service, handing a note to the chaplain in the pulpit. All soldiers were immediately ordered back to quarters, and in two hours they were on the road marching eastward, probably in support of the Union advance after the battle of Fair Oaks fought that day just east of Richmond. War had invaded worship, as it had invaded everything else in Virginia. But John heard yet more praying that Sunday; King's staff and aides left Warrenton "amidst the prayers and good will of the colored people that remained behind." Some of these "hundreds" of freedmen, their status still ambiguous, "followed . . . on foot. Poor mothers with their babys at their breasts, fathers with a few clothes in bundles . . . seeming to think this would be their sureist way to freedom."[50] One wonders if John was reminded of the slave coffles preparing to march southward that he had witnessed as a boy. Amid all this turbulence of war, black people in Virginia could now at least exercise their own will in moving about, and they had a new direction to march—north, or wherever the Union armies went. John Washington's story is a remarkable window into this tortured human flow toward freedom.

Within a few days Washington was back in Fredericksburg as part of General King's headquarters at the Farmer's Bank. This time he literally slept in his old room. He was reunited with his grandmother, his aunt (either Maria or Jane), and, though he does not mention it, his wife, who was now pregnant with their first child. "I surely was never so happy as then," he recalls.[51] As the war raged to a climactic moment only forty miles to the south, Washington lived with his liberators in one of the most distinctive edifices owned by his old masters. This was the best "hiring out" situation he had ever known.

In the savage fighting of the Battle of the Seven Days, the Federal army suffered sixteen thousand dead, wounded, and missing soldiers, and the Confederates approximately twenty thousand. But Richmond held, and soon General Robert E. Lee spearheaded a bold invasion of northern Virginia and Maryland. In sacrifice and purpose, the war was greatly expanding as it entered its second terrible year. The fate of John Washington, his family, and four million other slaves now depended directly on that expansion, as well as on the ever larger battles to come. On August 10, John again accompanied King's division as it speedily marched northwest toward Culpeper County. His view of the war was narrow, from the ground level of a general's staff headquarters on the move. On the previous day a fierce fight had occurred at Cedar Mountain just south of Culpeper Court House and twenty miles north of Gordonsville, a rail junction of major significance to the Confederates. The Lincoln administration now made the decision to move McClellan's large and nearly defeated army from the James River, south of Richmond, to Washington, D.C. In northern Virginia a Union army of fifty thousand, under the command of the blustering and inept John Pope, was expected to counter any Confederate forces threatening to move north. What Washington recorded in his narrative was his experience of daily skirmishes throughout the middle of August, a "campaign of thrust and parry," as one historian described it, between Pope's army and Lee's forces, now scattered and on the offensive, looking for an opening in order to threaten the Federal capital and transfer the ravages of war to northern territory.[52]

The emancipated aide-de-camp turned autobiographer punctuates his story with the details of stealing biscuits and hams from a bakery and a mill, and of a mission to find flasks of whiskey, personally ordered by General King. It would be many months before Washington could enlist under the authority of the Emancipation Proclamation as an actual soldier in the Union army. But like all soldiers, he could not forget the first time he came under fire. Accompanying King's division to Culpeper, in the midst of "air loaded with the thick perfume of clover and wild flowers and the heavy mountain dew looking like drops of silver on the rich leaves and blossoms," Washington recalls the "report of a rifle near by and the whistle of a 'Winnie' (Minnie) ball close to my head." Shocked out of gazing at the "beautiful mountain scenery" that continually captured his attention, John "dove the spurs" into his horse and quickly rejoined his unit. He appears to have ridden alone for some time along roads jammed with troops, cattle, artillery wagons, and plenty of "contrabands" on his mission to find the whiskey. He finally completed his journey, found King's staff, ducked as artillery shells hit the camp, and heard orders "whispered" since they were so close to "rebel pickets."[53]

Washington found plenty of real war, and he did his utmost as an unschooled writer to capture its drama. As he worked near the headquarters of Pope and the other generals, he knew "great preparations were being made for some important move." Soon the two armies would collide in the devastating Union defeat at Second Manassas that would take place on August 29 and 30, 1862. Washington, however, would not be there to see it. He received word that a reward of $300 had been offered for his "head" in Fredericksburg. Fearing for his family, he determined to go back to his hometown, retrieve his loved ones, and try to find permanent safety farther north. With General King's personal permission, he "bid farewell" to his companions and took the Orange and Alexandria Railroad from Culpeper to Washington, D.C., where he quickly sought a pass to Fredericksburg.[54]

Washington's formal leave-taking from the Union army that had helped so much to free him occurred sometime between August 14 and August 18. He departed just before Pope withdrew his army to north of the Rappahannock River and moved his headquarters to Culpeper. John was no longer serving as King's cook and aide when the general suffered a severe epileptic fit on August 23, seriously impairing his abilities and forcing him to yield command on the eve of the division's devastating defeat at the hands of Stonewall Jackson at Groveton on August 28.[55]

While John was traveling through Washington on his way to retrieve his family, President Abraham Lincoln invited a delegation of five black leaders to the White House on August 14 to explain his plans to settle African Americans in foreign lands. Many in Lincoln's administration were quite serious about spending the $600,000 Congress had allocated for such purposes. Lincoln himself, while privately crafting an executive order for emancipation and waiting in agony for some Union military success to provide an impetus to announce it, nevertheless worked publicly on behalf of these colonization schemes. The president even delivered a formal address to the black delegation that day in which he declared racial separation in America "a fact" and "equality" impossible.[56]

Lincoln did not conduct a discussion with the black delegation, he lectured them. The president's widely publicized statement could not have been better calculated to offend blacks at this hopeful juncture in the war. "You and we are different races," said Lincoln. "We have between us a broader difference than exists between almost any other two races . . . I think your race suffer very greatly, many of them living among us, while ours suffer from your presence." Thus, Lincoln concluded, "we should be separated." The president acknowledged that blacks were suffering "the greatest wrong inflicted on any people." But Lincoln blamed the war not only on slavery but on the presence of blacks. "But for your race among us," he insensitively declared, "there could not be war, although many men engaged on either side do not care for you one way or the other." With the five black leaders, who initially felt cordially welcomed, no doubt blinking in indignation, Lincoln asked them to take the lead as volunteers and consider expatriation to Central America. He sought a representative group "capable of thinking like white men," who would take on the task of settling a region with "rich coal mines" and good harbors.[57]

To some extent this ugliest of Lincoln's utterances on race may have been intended to condition white attitudes toward the revolutionary process of emancipation already falteringly underway. But that could hardly be solace to the Washington, D.C., black delegation; it must have hurt deeply to be told they had no birthright to equality in a nation tearing itself to pieces over their future. An anguished silence ensued as they were asked by the leader of the United States of America to "sacrifice something of your own comfort" to help the rest of their people make a new start in a faraway land.[58] As the excited John Washington sped through the nation's capital, unaware of this unfortunate White House meeting, he was one Virginian who did not entertain the slightest doubt about whether he belonged in the new country that might result from all the bloodshed and devastation in his home state.

On the steamer *Keyport*, Washington traveled down the Potomac River to Aquia Creek, where he disembarked and made his way slowly through Falmouth to the town of his birth for one last time. Finding Annie "as well as might be expected," he spent a week or so "enjoying my freedom with friends and acquaintances." John intended to spend as much time in Fredericksburg as possible while he and his wife awaited their firstborn, and he seems to have hoped for a permanent Union occupation. But the war rudely intruded after only a week of rest. As the Union forces began to abandon the city due to Lee's invasion northward, the former slave feared for his life. On August 31, Washington witnessed the Union forces dynamiting facilities and leaving Fredericksburg. He found an overwhelmed and terse provost-marshal, "saluted," and asked for a pass, which the officer begrudgingly issued. Leaving Annie in the care of women friends, and with fifty cents in his pocket, he crossed the wire bridge over the Rappahannock for the last time. He "looked back at the town that had given him birth and with a sad heart and full eyes thought of some of the joys I had felt within its limits." He "could not help weeping (though it was not manly)," as he thought of Annie, of vengeful Confederates, and, no doubt, of his entire youth spent struggling against the physical and emotional boundaries of slavery on the banks of that river.[59]

Washington was a fugitive once again. Trekking through a countryside no longer protected by Union troops, he hid out in "thick undergrowth" and worried about which army he would encounter on his hike to Aquia Creek Landing. He found his way to the landing and the same vessel he had boarded two weeks earlier. But this time his passes as a member of General King's staff were rejected, forcing him to stow away on the *Keyport* by bounding "across the gang plank" when the guard was not looking. Soon he arrived back at the Sixth Street Wharf in Washington on what he remembered as the night of September 1, 1862. Pope had just ordered his beaten but not routed army to pull back within the entrenchments and streets of the now threatened capital city.[60]

Washington claims that his grandmother, Molly, as well as his aunt Maria and her four children, arrived in the city with him. In all likelihood they came separately, but he does not tell us exactly how. For one night, he reports, they all "slept on 14th Street," which runs just one block east of the White House grounds. The next morning, dodging horses, caissons, and straggling soldiers, they all walked at least three miles into Georgetown, seeking "some place to stay."[61] But John spares us the harrowing details of his personal band of refugees looking for shelter and food in the war-enveloped city.

Washington and his desperate kinfolk were not alone; contraband camps had sprung up all around the District of Columbia in the midst of military crisis and disaster. The family somehow found a place to board at $2.50 per week while Washington looked for work. He confessed he was "not very strong" and desperately sought any "light" labor, remarkable statements in an autobiography that normally asserts his own heroic actions. He probably gave what food he could obtain to his family. His formal narrative ends with the simple declaration that he "finally obtained a place bottling liquor for Dodge & c at $1.75 per day, which lasted for some time."[62]

With great hardship and severe dislocation, John Washington and his extended family had begun a new life. On October 6 Annie gave birth in Fredericksburg to William Herbert Washington, and she joined John and the other relatives soon thereafter by undetermined means through war-torn northern Virginia. More blood relatives would later follow, including John's mother.

In the period between Lincoln's preliminary and final emancipation proclamations, Washington's extended family made the best of bad conditions. In a way, they did become colonists, but not to a tropical shore in Central America; their colony of choice took shape in the shadows of the White House and the United States Capitol.

CHAPTER TWO

Mobile Bay

> It was death to go back and it was death to stay there and
> freedom was before me; it could only be death to go
> forward if I was caught and freedom if I escaped.
> —WALLACE TURNAGE, REMEMBERING MOBILE BAY,
> NEAR DAUPHIN ISLAND, ALABAMA, AUGUST 1864

Wallace Turnage was born on August 24, 1846, in the Tyson's Marsh district of Green County near Snow Hill, North Carolina. He was the son of a slave woman named Courtney and a white man, Sylvester Brown Turnage, who was the stepson of Courtney's owner, Levin Turnage. Unlike John Washington, Turnage knew the identity of his father and listed his name on all manner of official documents for the rest of his life. How Courtney and Sylvester Turnage came into sexual contact in such a small agricultural setting is not hard to imagine. Sexual abuse of slave women by young white men on farms and plantations rarely happened in random attacks; it occurred as a result of white control over a black woman's time, labor, movement, and body. When a slave girl reached her teenage years she was vulnerable to rape or sexual intimidation. Courtney was about fifteen and Sylvester Turnage was eighteen when she gave birth to Wallace. He became one of the nearly quarter of a million slave children who were of mixed race in the 1850s.[1]

Levin and his wife, Senetty Turnage, owned only four slaves. Levin died in 1836 when Wallace's mother was about five years old. Senetty remarried Hewell Hart of Snow Hill in 1837, and it was on their cotton and tobacco farm where Wallace began his life as a slave. Sylvester Turnage married in 1850, when Wallace was four, and moved to another county. Even if Wallace's father had paid any attention to him or acknowledged his paternity, the young slave probably saw little of him after that date. Wallace's mother eventually went by the name of Hart; she married a Louis Hart, probably a slave, sometime in the 1850s, and had four more children. When Hewell Hart died in 1855, Senetty became the sole owner of Wallace, his mother, and his siblings. And by 1860, when Wallace begins his narrative, Senetty Hart owned a farm worth $1,800 and seven slaves.[2]

As the election year of 1860 brought Abraham Lincoln and the antislavery Republican party to power and the crisis over slavery's future swept the nation, Wallace Turnage's world of the small farm in central North Carolina suddenly expanded both geographically and psychologically. In this remote yet complex familial setting, where the boundaries of race and sex were fluid and abusive, Wallace was sold away by his indebted owner to a Richmond slave trader, Hector Davis. Turnage arrived in Virginia in the late winter or early spring of 1860, a scared, still-growing fourteen-year-old, purchased for the robust price of $950 and cut off from any moorings of family or home. We are not told of a parting with his mother in Snow Hill, but from that day forward Wallace could only understand his fate as that of an orphaned teenage slave forced to survive in a dangerous and tyrannical world by himself. The Civil War would soon open a passage into hell for Turnage, but it would also lead to his ultimate liberation. In the opening lines of his narrative, he promises a sketch of his "adventures and persecutions."[3] For the next four years he had plenty of both.

By the 1850s Richmond had become a major supply center for the domestic slave trade, a depot for reselling slaves to the deep South, and a magnet for planters who came shopping for field hands and domestic servants. When Turnage arrived in 1860, Davis's auction house and jail was thriving as one of the three largest such establishments in this city of some 38,000 inhabitants, 14,000 of whom were black. By one estimate from 1858, slave sales in Richmond netted $4 million that year alone (approximately $70 million in 2007 dollars). Slaves were sold both at public auction and by private arrangement. The city's slave traders were the middlemen in a long-distance trade in people that was a major cog in the local and Southern economy. One 1858 market guide for slave sales in Richmond lists average prices for "likely ploughboys," ages twelve to fourteen, at $850 to $1,050, "extra number 1 field-girls" at $1,300 to $1,350, and "extra number 1 men" at $1,500.[4]

On any given day, Davis's auction house would offer as many as seventy to one hundred slaves for sale. In the Richmond Business Directory, Davis advertised himself as "Auctioneer & Commission Merchant for Sale of Negroes." His account book reported "cash for negroes this week" on October 27, 1860, as $14,710; and a month earlier, Davis recorded one sale of ninety-five slaves to N. M. Lee for an astounding $135,785.50. Some smaller traders tried to disguise this part of their business interests, but not Davis. His promotions announced that the proprietor "sells negroes both publicly and privately and pledges his best efforts to obtain highest market prices." He further offered a "safe and commodious jail where he will board all negroes intended for his sale at 30 cents per day."[5]

The auction house was a three-story building with a bell tower, two balconies inside, and, according to an eyewitness who observed its remains for the Works Progress Administration in 1937, "beautiful arches" with "wooden grill work." The walls were made of heavy stone, and the floor measured approximately one hundred by fifty feet. Immediately next door and across the street were the Exchange and Ballard hotels, where planters, traders, merchants, and brokers of all kinds boarded and conducted business. Here within a three-block radius of Franklin and Fourteenth streets were several churches, more hotels, and the Odd Fellows' Hall. This was the business hub for Richmond's fifty or so tobacco factories, its various other commercial enterprises, and the slave trade. Not confined to auction houses alone, slave traders sold people in all manner of buildings, taverns, hotels, and even in the post office, which was housed in the Exchange Hotel.[6] All of this commerce ensued just a few blocks down the hill from the stately Virginia capitol building, where within a year of Turnage's arrival at Davis's slave jail the Confederate Congress would meet.

It was in this dark and ugly crossroads of American slavery that Turnage had landed. Here he remembers laboring by day as Davis's "auction and office boy . . . taking people from the jail to the dressing room and from the dressing room to the auction room." Turnage avoids telling us much about the horrors he must have witnessed working in Davis's slave pen—the leg irons on slaves, the iron rings in the floor, the flogging of slaves with broad cowhide paddles so as not to leave too many scars before sale, the cries of physical pain and human separations, and the body in-

spections. The fourteen year old would have seen and heard all
the methods by which slaves were graded or categorized for
sale—"fancy girls," "prime" and "likely boys," "buying by the
pound," "bright mulattos," and "jet black negroes." He must have
seen slaves forced to dance or sing before a sale, their hair dyed
and gray whiskers plucked to mask their age. Turnage witnessed
the myriad ways bodies were displayed for purchase; he watched
as slaveholders imagined their fantasies and fortunes in the people
they bought and the lives they stole. In his 1848 narrative, former
slave William Wells Brown, born in Kentucky but sold to the
New Orleans slave market, describes doing the same job as Turnage—assisting the sale process by helping slaves look cheerful
and ready. "Before the slaves were exhibited for sale they were
dressed and drawn out into the yard. Some were set to dancing,
some to jumping, some to singing, and some to playing cards . . .
My business was to see that they were placed in those situations
before the arrival of the purchasers, and I have often set them to
dancing when their cheeks were wet with tears."[7]

Turnage witnessed all of this "business," and soon he was up
for sale himself. In the late spring of 1860, Davis sold Turnage for
$1,000 to Scottish-born James Chalmers, a cotton planter who
came to Richmond from Alabama twice a year to buy slaves. Turnage's remembrance of this exchange offers a fascinating variation
on the corrupt language of slave buying. One morning while
walking up Franklin Street on an errand, Turnage was whisked
back into the slave jail and told that he would be among the
"drove" transported the next day. According to Turnage, Chalmers
was a "very clever man" who told him he lived nearby as he "asked"
the teenager if he "would like to live with him." At 4 A.M. the following morning the young slave was on a train to Petersburg,

Virginia, and from there two days later he arrived in the tiny town of Pickensville in west-central Alabama near the Mississippi line.[8]

Wallace Turnage had been swept into the massive American domestic slave trade, the scale of which he could hardly imagine. As the production of cotton, America's greatest export, spread from the southern seaboard to the Southwest, approximately one million slaves were carried into Alabama, Mississippi, Louisiana, and Texas between 1820 and 1860. Two-thirds of these slaves went by outright sale through traders. The numbers tell a tale of cold speculation by slaveholders unloading excess assets from the upper South and coastal regions, where production declined, to the feverishly expanding Cotton Kingdom: 155,000 in the 1820s; 288,000 in the 1830s; 189,000 in the 1840s; and 250,000 in the 1850s. One recent study concludes that in 1859 and 1860, the value of slaves sold in the domestic market was $9.26 million. Indeed, the value of American slaves as a whole was the single largest asset in the entire national economy, larger than all the railroads and manufacturing firms combined.[9]

In the four decades before the Civil War, the transfer of a half billion dollars in human property across the landscape fueled the economy of the whole nation, not merely that of the South.[10] Many slave owners and traders insured slaves (often with Northern companies and banks) while they were in transit and bought life insurance policies after their arrival in the Southwest. This was a complex, thoroughly legal business, with many depots, lending banks, brokers, middlemen, insurers, and subsidiary firms. It was protected by government and fueled by private enterprise; it was capitalism with no apologies and ever-deepening racist justifications. When cotton coughed, the economy got a cold in antebellum America; but the deeper disease festered in Hector Davis's slave jail and many others like it—in the hungry quest for fortunes made with black bodies and the labor they could perform.

As he arrived on a cotton plantation in Alabama, the teenage Turnage felt a terrified bewilderment over how he could be both so valuable and so brutalized at once. On his first day he was, by his account, quickly "initiated according to their way of treatment." Welcomed by the other slaves on a work crew who were eager to hear any news of "Old Virginia" (the home from which many of them had been sold away), Turnage soon met a "shabbily dressed man" with a bull whip—his first of many cruel overseers. The overseer pulled a woman out of the group for some offense, yanked her dress up, and whipped her "shameful." Turnage, who had spent his youth on a tobacco farm where the treatment was less harsh, declared this scene "rough for the first day."[11] He was now a true field hand.

That night in the quarters, the disoriented Turnage seemed astonished as his fellow slaves engaged in "fiddling and dancing as though they had no oppression at all." Wallace spent the summer trying to acclimate to the dawn-to-dusk labor and recurring violence. Horns blew in the morning to awaken the weary laborers. Slaves were frequently beaten, especially women who were forced to "strip right before the men." By the fall Turnage had himself been beaten three or four times, and he had resolved to "go in the woods and go back where I come from."[12]

Despite his fellow slaves' dire warnings that he had no hope of escape, Turnage ran away for the first time that fall of 1860. For most of his ensuing narrative, he struggles to create the persona of the heroic runaway—a young, desperate, tortured rebel confronting an evil system and calculated violence. Irrationally, but persistently, he wanted to run away "home," wherever and whatever that actually meant to him. Wallace had no home except for his tangled memories of Snow Hill and his mother. Whether those memories sustained him we cannot know. His faith that "the Lord" somehow protected him in his times of hunger, fear, and physical pain clearly buttressed his spirit. But Turnage found his ultimate freedom through a tunnel of horrors.

As autumn arrived, "about corn gathering time," he experienced "sore hands" and could not pick enough cotton to meet the overseer's demands. Two slave women also fell behind on picking their prescribed allotments. Turnage describes a sadistic overseer who forced the two women to the ground and delivered "two hundred lashes" to one of the women's naked back: "I could see the skin fly about every lick he struck her." Rather than wait for his beating for failing to pick enough cotton, he slipped off into the woods.[13]

Turnage admits he "didn't know where to go" on his first escape and had to return to his plantation when he was overtaken by hunger. Other slaves claimed the master would protect the famished fifteen-year-old from the overseer's whip if he would come home, and, in fact, his master did. His owner kept him safely at the big house for two days before returning him to field labor. Horton, the overseer, "pretended" to be glad about Wallace's return; but Turnage imagines scenes reminiscent of Frederick Douglass's famous struggle against his overseer, Edward Covey, who had been hired to tame Douglass's rebellious ways. Covey's "forte," wrote Douglass, "consisted in his power to deceive."[14]

Turnage now engaged in a war of wills with his deceptive tormentor. As Horton approached him with the cowhide in hand, Wallace determined to make a stand and fight. He refused to obey any commands and "spoke very saucy." With a bravado typical in many male slave narratives, Turnage declares himself an "expert wrestler," and he took on the overseer in a two-hour brawl in which his savage foe bit his ear. Wallace got the best of the fiendish Horton, but eventually the white man enlisted a strong male slave from among the thirty or so watching the fight to tie the young rebel down. Held to the ground, hands tied to a tree, Turnage endured ninety-five lashes (his count) as the price of his resistance. As a writer, he could not avoid using the episode for a little comic relief: He describes a scene in which Horton visits a fellow overseer that evening and embarrassingly must account for all the scratches on his face from his rumble with Turnage. By reconstructing a dialogue between the two agents of slavery's violence, the former slave relishes his remembrance of "marking" the cruel overseer's face.[15]

Turnage attempted a second escape less than a year later, prompted again by the overseer's threat of more beatings in response to his insubordination. By June 1861, from the slave grapevine of information as well as the military mobilization he could see around him, Turnage had gained two advantages—a much better grasp of local geography and the knowledge that the war might soon come to Tennessee and Mississippi. Turnage lit out across the Alabama line into Mississippi for what he hoped was a "refuge in a strange country." He traveled as far into the state as Columbus (about twenty-seven miles from Pickensville) before he encountered a white man with a shotgun who fired and missed. Turnage writes of these escapes as both adventure and horror. Bullets flew by his "ears and shoulders cutting the leaves like so many hornets," but he outran them. He feared only men with horses or guns, he maintained, because he was a "very fast runner." In his "wild and half-starved" condition, Wallace appears to have circled back toward Pickensville.[16]

There a white man along the road challenged his identity. Slave codes required that a slave carry a pass when away from his plantation. Not able to produce one, Turnage jumped a fence and ran into a cornfield at such a speed that the "roasting ears almost knocked me down." He burst onto a group of slaves working in a field, with the master's dogs nearby. Some of the bondsmen threw their hoes at him and the dogs set off in hot pursuit. Turnage ran until he came upon a "very deep ditch." "The Lord" helped in his desperate leap over the ditch, and the dogs fell in and could not get out.[17]

Once Turnage arrived back in Pickensville, a friend gave him food and hid him in the steeple of his own master's church. The torrid heat and lack of water brought him down every night to get food from his protector. By day Wallace could see his master "as he passed by . . . going to his store but he did not know that I was there." Turnage's brief residence in the steeple, observing the town and the slave economy below, is not unlike Harriet Jacobs's much longer and more famous years of confinement in her grandmother's attic, peering through her "little loophole," where she "heard the patrols and slave-hunters conferring together about the capture of runaways." There she "peeped at [her] children's faces, and heard their sweet voices." Wallace's travail in the hot steeple was not nearly as anguished nor as long, but he, too, yearned to "breathe the pure air," as Jacobs had. That yearning might have been what drove him outside during the day, where he was caught by a man with a pistol who saw the young fugitive hiding in some bushes, his white straw hat too obvious to miss.[18]

Frustrated, but not willing to give up his valuable slave, Chalmers now kept the teenager close to him as a house servant. But not without beating him severely. Even Chalmers's wife had Turnage whipped until the slave-owning couple "thought they had broke me." But by late November of 1861, "tired of being whipped," the intrepid sixteen-year-old bolted up the road toward northern Mississippi.[19] On this third attempt, he would make it much farther and stay at large until the depth of winter on a journey he portrayed as both thrilling and terrifying.

Indeed, Turnage's memoir fits a genre critics have called the fugitive slave "road narrative." Henry Bibb, in his 1849 *Narrative of the Life and Adventures of Henry Bibb*, describes his five different escapes and captures between 1837 and 1841. The thing he knew best, he said, was the "art of running away," and from an early age, he "made a regular business of it."[20] Turnage's third escape saga is replete with episodes of extreme hunger, high rivers to cross, military guards at bridges, and hiding in barns, a gin house, and a church. Some fellow slaves take him in and protect him, and at least a couple do not.

Unlike some other road narrative authors, Turnage never expresses any moral anguish over the lies, deception, theft, and breaking down of doors that enabled him to survive. Heading north toward Tupelo, Mississippi, where he could at least hope to eventually reach the Union armies, Turnage struggled against bloodhounds, the winter cold, near starvation, and extreme brutality until he was finally captured. He portrays himself as an agile youth with desperate courage and some good luck, but who was destined to be caught by the police state of slavery in wartime Mississippi.

When Turnage sat down to write his narrative, he obviously possessed an extraordinarily accurate and detailed sense of geography. Perhaps he consulted maps or read war histories to refresh his memory of the terrain over which he had journeyed. Turnage names the towns or railroad crossings in the order of his route, all of which can be found on any modern map. After leaving Columbus he crossed the Luxapatilla River, and on the road to Aberdeen he negotiated the Tombigby River in a stolen rowboat after trying to ride a log over the swollen stream. From Okolona, where he delicately circled around a Confederate army encampment, to Egypt Station, where he hid in a church tower, and on to Shannon ("shanel") Station, just south of Tupelo, the determined runaway traveled at least seventy-five miles in more than a month before he asked for food at one too many slave cabins. There a frightened black woman turned him in to authorities after her dog cornered Wallace inside her cabin. Some slaves feared runaways because they often brought disorder, angry masters, and punishment wherever they appeared.[21]

Captured, interrogated, and tortured by white men who did not believe that he could hail from so far away, Wallace was pistol-whipped and stabbed through his thick coat. One of his sadistic interrogators smashed his head repeatedly into the bricks under the mantel of a fireplace and threw the slave into the fire as though to kill him. Wallace scrambled out, his hands "badly burnt." He was "locked around the neck" and "chained" to another fugitive slave for the night. Eventually a third runaway was added to the group, and they suffered in mutual agony and filth while locked in the same chain for another five days. Oddly enough, Turnage reports that the brutal white man "repented of what he had done to me" and found a key to unlock the chain.

The burned and bloodied youth was taken by handcar down the railroad back into Alabama. Chalmers paid the man fifty dollars for the return of his property, delivered twenty-five lashes to his fugitive's back, told him he would "never get back" to North Carolina, and sent him out to the plantation.[22]

With a sense of drama and tension, Turnage wants his readers to know that his third escape almost succeeded. He stresses that he was never taken without a fight and that he never surrendered to his dehumanization. Turnage describes holding on desperately to bushes and reeds as he scrambled along a river bank that rose one hundred feet high while a white man pursued him: "I had to go very fast, and it seemed to me that every particle of ground that I took my foot off of on this bank rolled down in the channel of the river. I heard the man hollowing [sic] after me but I kept on and got by all safe." In that dirt crumbling into the watery abyss under his feet as he ran, Turnage struggles for a metaphor for his own condition. That night he found refuge in a barn at a large plantation, trading "sugar for meat" with local slaves and sleeping long into the next morning.[23] But nowhere on this journey did Wallace find either solid ground or true safety.

In an environment full of danger, betrayal, and the slow, enveloping beast of war, Turnage portrays himself as the lone heroic fugitive slaying dragons and fighting the dark odds, aided by "friends" until he simply runs out of options, routes, or comrades. He becomes a soldier of a kind, fighting his private war against slavery, but also a prisoner in that war, constantly breaking out and being recaptured with ever-increasing violence. Wallace Turnage fought a war within a war well before he ever saw a Yankee soldier.

Turnage's narrative is a fascinating illustration of the unorganized networks of slaves who clandestinely aided runaways. It even allowed the former fugitive to suggest his longstanding yearning for a home. At one point an old black woman hid him from her suspicious neighbors by telling them that Wallace was her son returning from a long journey. And an "old colored gentleman" took him into his house, fed him, and protected him for a night.[24] Turnage's Mississippi escapes show a variation of the Underground Railroad as it functioned in the deep South in wartime, a term fraught with modern mythology but rooted historically in a process of blacks aiding the escapes of other blacks.

In some of the details of Turnage's story we may find clues for why he wrote the narrative. He clearly wanted his travail known to his family and friends. He wanted respect for his endurance of the hell through which he had passed. His war story might gain an acknowledgment that an otherwise ordinary and unknown black man in the struggling postwar working class was truly somebody. His escapes provided badges of honor: his ever-present, intense hunger; his blood drawn by whips, chains, and torture; his ingenuity, trickery, and deception at almost every turn in the road; his skills for sheer survival (baking his own bread after breaking down a kitchen door); his teaching an older slave man to read in exchange for food; and the remarkable recognition that he received during the third escape from the five male slaves who hid and fed him for a night on the road to Aberdeen. Feeling "only a boy" among these men, he writes in the voice of a lonely youth taken in by elders who protected him. "They was all my friends . . . they gloried in my spunk."[25] And then the hero ran on.

The story of Turnage's many escape attempts is deeply intertwined with the war itself. Fearless, with nothing to lose but his increasingly scarred body, he kept trying to reach the war front in northern Mississippi or Tennessee and there find a path "to get to the free states." In 1862 the war had enveloped the northern tier of Mississippi and brought chaos to the countryside, threatening and dissolving the institution of slavery in some areas. The war pulled Wallace into its irresistible grip, but his fate depended on his courage to seize his freedom. His third escape attempt came as real fighting occurred in Kentucky, Missouri, and Tennessee, as well as in key places along the Mississippi River. All around him white Alabamans and Mississippians by the thousands were marching to war, leaving women and slaves to occupy the home-front. By the time of his fourth escape attempt in the fall of 1862, he was entering a real war zone in the immediate wake of the battles of Iuka and Corinth, the costly struggles between Union and Confederate forces for the control of northern Mississippi and western Tennessee.[26]

This was also the period when the issue of emancipation had begun to transform the purpose of the war. Turnage hardly could have been aware of the policy struggles over emancipation within both governments, but he surely knew that the awful war engulf-ing the Southern heartland offered the best chance a teenage run-away would ever see. Turnage's narrative provides a remarkable example of the alternately naïve and informed bravery that it took for many slaves to escape in the midst of such chaos. At the same time it also demonstrates the centrality of the war in the prospects for any such escape. Turnage's story is one desperate collision after another with the savagery of both slavery and war, and he ulti-mately stole his freedom from the clutches of both.

Turnage was hardly a lone slave refugee. Congress authorized the freeing of slaves owned by disloyal masters in the Second

Confiscation Act in July 1862, and Lincoln issued his long-contemplated Preliminary Emancipation Proclamation that liberated slaves in the "states in rebellion" on September 22, 1862.[27] So by the fall of 1862, northern Mississippi had swelled with runaways seeking whatever protection, employment, or hope they could find among the Union forces. It was common for Yankee officers to send dispatches describing "a Negro just in" who reported Confederate troop size and movements. But Union commanders were divided on just what to do with the growing numbers of "contrabands," the term used since early in the war for escaped slaves entering Federal lines.

Some military officers welcomed them; generals Ulysses S. Grant and William T. Sherman eagerly sought black laborers to build fortifications and do all manner of other military work in the region. "How many contrabands can you furnish for work on the fortifications at Corinth," Grant wrote to one of his generals. And Sherman, commanding the Union occupation of Memphis, frequently reported his rising number of black laborers. "We have about 6,000 negroes here," he wrote on October 29, 1862, "of which 2,000 are men—800 on the fort, 240 in the quartermaster's department, and about 1,000 as cooks, teamsters, and servants in the regiments."[28] At the very time Grant and Sherman wrote their dispatches pleading for ex-slave laborers, Turnage was either hiding out along the Mobile and Ohio railroad line or in custody on his way back to the Alabama cotton fields. On Turnage's fourth flight he nearly made it past the Confederate camps around Tupelo to the Union lines at Corinth in the northeast corner of Mississippi. He would have been happy, at least for a while, to help construct the elaborate trench works with which the Yankee armies surrounded Corinth.

As the fall turned to winter, Grant and many of his commanders worried over what to do with the growing numbers of escaped slaves sweeping into their ranks. "Citizens south of us are leaving their homes and negroes coming in in wagon loads," Grant wrote to General-in-Chief Henry Halleck in November 1862. "What will I do with them?" And again, in early January 1863, he queried Halleck: "Contraband question becoming a serious one. What will I do with surplus negroes?"29 The scale of Grant's problem would only grow as he launched the campaign to take Vicksburg on the Mississippi River that spring and summer.

For their part, Confederate commanders did their best to discourage slaves from escaping all over the cotton-belt region. A "General Orders" issued by Brigadier General Daniel Ruggles in July 1862 declared an unequivocal if wishful Confederate policy: "The passing of negroes, slaves or free, toward or into the enemy's lines is positively prohibited without a pass from the master . . . countersigned . . . by the provost-marshal or highest military authority." Reflecting a sense of desperation over a reality that the Confederate forces no longer controlled, the order tried to leave no room for doubt: "Every negro, slave or free, who shall violate this order will be shot in the attempt to pass the lines or beyond the prescribed limits indicated by our advanced pickets or guards."30 It was just those limits that Turnage tested over and over again, and it is no small wonder that he was not killed in the effort.

Although he seems to have run alone, Turnage was part of this growing wave of fugitives causing fits for Confederate commanders in the deep South. From a Southern cavalry headquarters in Okolona in January 1863, an adjutant wrote to his superiors complaining about captured runaway slaves who refused to give the correct "names of their owners and residence." Their swelling numbers were "increasing beyond convenience," he wrote, especially since a January 1 emancipation ceremony in Corinth that was conducted by Union chaplains and officers in conjunction with freedmen and reported by one fugitive, at which slaves were given pistols and told "they were free."[31] Turnage would have many miles to go before he could experience his own emancipation ceremony. Meantime, he did his utmost to be an inconvenience for the Confederacy.

Rife with troops, foraging stragglers, and destitute civilians, the countryside through which Turnage fled had smaller plantations and fewer slaves than did the heart of the cotton belt. But it had two important railroads. The east-west Memphis and Charleston and the north-south Mobile and Ohio railroads intersected at Corinth and gave birth to the town only five years before the war. In 1861 Corinth boasted 2,800 inhabitants, some fine houses with Doric columns, and the Corona Female College. The Tishomingo Hotel aspired to some splendor, but the swampy area had bad water, and it became especially unhealthy when fifteen thousand of the eighty thousand Confederate troops headquartered there got sick in the summer of 1862. A Confederate colonel judged Corinth a "sickly, malarial spot fit only for alligators and snakes." And with even less charity, a Minnesota lieutenant called the local men "ignorant" and the women "she-vipers," with the shapes of "shad-bellied bean poles." The "principle products" of the area, concluded the Yankee, "are wood ticks, chiggers, fleas, and niggers."[32] From his experience, Turnage might have added an even worse characterization of the region, but the Union-occupied Corinth of 1862 and the Mobile and Ohio Railroad were the only pathways to his dream.

———

From February to August of 1862, Turnage labored on the Chalmers plantation under the strictures of a new overseer named Scoggen. The incorrigible runaway and the wary overseer engaged in the customary psychological warfare. Turnage was never a very pliant or effective cotton slave. Scoggen whipped Turnage only once, but it was a severe beating with a "walking stick," and the former slave remembers that "he hurt me very much . . . I took it but it went very hard with me." When Scoggen threatened him with another beating, Turnage fled once again down the road toward Columbus. He mocks the hapless overseer, claiming that he "had forgot that I was such a soon runaway."[33]

From some combination of courage born of both hatred and perhaps what modern analysts call trauma, Turnage made a beeline for the Yankee armies that he knew were concentrated around Corinth. Sick from eating too much fruit, "completely exhausted" from what he claimed were six nights without sleep, Turnage found help just south of Okolona, where a black man with a rowboat ferried him across the Tombigby River. The boatman understood "my condition," writes Turnage, and "asked me no questions." A "friend" from a previous flight provided food and directions to Corinth. Wallace now traversed what historian Stephen Ash has characterized as the three "zones," or "worlds of the occupied South"—the "Confederate frontier," "no-man's land," and "garrisoned towns." In Turnage's story, the frontier consisted of the hinterland counties of northern Mississippi, where few Yankee patrols ever penetrated and Confederate home guards or regular troops generally controlled the public roads. The no-man's land was the region closer to the garrisoned Corinth where civilians had fled and only troops and guerrillas trod. At one point near Okolona, the runaway finally had to stop and sleep. Somehow he survived by hiding in a fence corner for nearly two days as he watched Confederate troops pass.[34] It was now Turnage's war, too, whatever the consequences. It is clear from his sheer persistence that he understood this vast cauldron of war to be somehow about whether a teenage cotton slave had the right to possess his own life. He wanted to make those Yankees his liberators.

Since he was in the no-man's land of a war zone, Turnage walked "boldly upon the railroad" until he reached the outskirts of the town of Rienzi, just south of Corinth. There he got "careless," and encountered a "rebel scout" on horseback who chased him into the woods. Suddenly surrounded by a band of Confederate cavalry (possibly guerrillas or home guards), with a pistol pointed at his head, and confronted by a man with a bull whip slung around his neck, the experienced runaway was mauled by dogs for what he described as four to five minutes. He was held by the Confederate scout at a house in the vicinity for two months (late August to mid-October of 1862) while the white man sought recompense by repeatedly writing to Turnage's owner in Alabama.[35] The desperate and wounded fugitive spent his seventeenth birthday in lonely captivity at a bounty hunter's house, half a day's walk from Union army lines and possibly within earshot of the two-day battle of Corinth where 828 men were killed and 2,800 were wounded.

Wallace's master finally risked life and limb to come to this remote northeastern Mississippi farmstead to retrieve his valuable property. Traveling with only a mule at first, Chalmers and Turnage proceeded southward to the Confederate interior. On one overnight stop the runaway was chained to his master's bedstead to prevent yet another escape. Once again brought by a handcar to Okolona, then by the railroad to Columbus, and finally by a rented wagon to Pickensville, Wallace Turnage returned for a fourth time to the Cotton Belt, a failed but veteran runaway. He and his master confronted some rebel pickets not far from home, who inquired as to how Chalmers had obtained Turnage. When hearing the story of Wallace's incessant escapes, they demanded to take the young man, tie him to a tree, and use his body "for a target."[36] Chalmers protested that his slave was too valuable and that he could still sell him. The men backed down, but not before terrifying the captive to their delight. Although Turnage lacks literary polish, his metaphors represent the full range of nightmares in the white Southern imagination: the hunted young boy cornered and bitten by dogs, the valuable bondsman chained to his owner's bedstead, and the dangerous and untrustworthy slave fit only to be used as target practice. Turnage's war was all about his freedom, as he was reminded by both those who would save him for money and those who would kill him for sport.

Chalmers wasted little time in trying to sell Turnage. After just a few days the planter took Wallace by overnight train to a slave trader's yard in Mobile, leaving him there to be sold to the highest bidder. One trader demurred after learning of Turnage's history as a runaway, but another put him to work in a store while he waited for a buyer. Wallace now became an urban slave in a port city transformed by the war. Sometime in December 1862 or January 1863 he was finally sold privately (not at auction) to an "old gentleman," Collier Harrison Minge, a prominent merchant and the nephew of former president William Henry Harrison. Minge needed a carriage driver and a house slave, but he got more than he bargained for when he bought Turnage for $2,000 in Confederate money.[37]

Turnage could see that living in Mobile might improve his chances of escape despite its being a well-fortified city and garrisoned by some ten thousand Confederate troops. In time the war would come to him instead of his having to find it. General William T. Sherman's Meridian Expedition of January and February 1864 cut a devastating path across Mississippi, destroying bridges, rail lines, and locomotives, and spread the fear of a federal assault on Mobile that never materialized. Confederate general Nathan Bedford Forrest's success in driving back Union forces in northern Mississippi left Mobile the prey largely of a Union naval fleet in the Gulf of Mexico.[38] Sherman's Federals backtracked to Vicksburg and left most of Turnage's earlier escape routes without a Union presence. Had Turnage remained on the plantation in Alabama, the lines of the liberating Yankees would have been farther from his grasp. As the war raged and Turnage drove his new master's carriage around Mobile, the young slave learned much from his fellow blacks in the city as he waited, plotted, and survived.

For a slave from the interior of the Cotton Belt, Mobile must have seemed a bewildering, cosmopolitan, even exotic place at first. A city of approximately 30,000 people, 7,500 of whom were slaves and nearly 1,200 of whom were free blacks, Mobile was the South's second-largest seaport after New Orleans, as well as a major railroad and river transportation junction. Many among the free black population were Creoles, people of mixed racial heritage (French or Spanish and African) whose ancestors had been free during the colonial period and who had enjoyed some citizenship rights under the 1819 treaty that ceded West Florida to the United States. Some Creoles were loyal to the Confederacy and tried to enlist in the army, even organizing "Creole fairs" to benefit the cause. Most of the slaves worked the city's dozens of cotton presses, as draymen, as domestic labor, or in the warehouses on the wharves. A British journalist visiting Mobile in 1861 observed the marketplace "crowded with Negroes, mulattoes, quadroons, and mestizos of all sorts, Spanish, Italian, and French, speaking their own tongues, or a quaint *lingua franca*, and dressed in very striking and pretty costumes." Significant numbers of Irish and German workers had arrived in Mobile on the eve of the war as well.[39] Turnage's frail, light-skinned visage must have blended well with the local population.

The city that rested on a delta shaped by multiple river channels was threatened from the sea by Union warships, and it suffered material shortages and at least one bread riot by women in 1863. But it remained relatively unscathed by war itself, and many travelers were stunned by its wartime beauty. "What a beautiful city Mobile is, quite a feast for the eye," wrote Frances Woolfolk Wallace in April 1864. "I think it the most beautiful place I have ever seen." This Southern lady, the wife of a Confederate officer she had followed on campaign, found the whole city "perfumed" with spring flowers as she strolled its streets—no doubt a pleasant contrast to the "scene of destruction and desolation" and the "land . . . flowing with human blood, death and suffering" that she had witnessed in Mississippi.[40] From an entirely different perspective—his "visit" began in a slave pen—Turnage might have agreed with Mrs. Wallace, as he, too, smelled the honeysuckle, observed the mansions bedecked with roses on Government Street, and marveled at the stately live oaks of Mobile that spring of 1864. And he, too, liked gardens.

Turnage seemed to grudgingly respect his new owner, who wanted him to tend the house garden. Gardening seemed to

please Turnage, perhaps because he garnered some joy in cultivat-
ing vegetables and spring flowers, but more likely because "I knew
all about it." Minge's wife and daughters wanted his services in
the house, too, and Wallace spent his days responding to everyone's beck and call. The young slave joined a Baptist church, although he lamented that the Confederate troops' unusual control
over the Sunday activities of slaves kept him from experiencing a
formal baptism in a pond or river.[41]

The war inspired Mobile city officials to pass even more restrictive laws to preserve slavery. The City Code prohibited slaves
from renting out their own time and from owning dogs and cows
or even from smoking in the streets. Those slaves allowed to hire
out by day were required to wear a metal tag issued by the city.
(John Washington would have faced more trouble in Mobile than
he did in Fredericksburg.) In 1863 the Mobile Committee of
Safety even petitioned the governor to ban publication of the
Emancipation Proclamation.[42] As a teenage veteran of slavery's
chains and whips, Turnage no doubt ignored at least some of
these legal precursors to Jim Crow.

The Minges' carriage had an old, worn harness for the horse.
On a hot day in early August 1864, the harness broke in front of
the stores on Dauphin Street. As the horse bolted away Turnage
was thrown to the ground, narrowly avoiding a streetcar. Soldiers
retrieved the horse and put Wallace on its back for the ride home.
This was perhaps the only kindness he had ever experienced from
Confederate troops and the last humane interaction with white
people he would have in Mobile. As his mistress excoriated him
for the broken carriage, Wallace "got angry and spoke very short
to her." Aware that his reaction was "counted the height of imper-

tinence of a slave in the South," Turnage expected it when Mrs. Minge ordered him to the dank cellar.[43]

In all likelihood, Turnage's conditions at the Minges' house were the best he had known in his anguished young life. His church might have given him meaningful companionship for the first time, and the urban cosmopolitanism of wartime Mobile undoubtedly fascinated him. Moreover, even a slave with Turnage's keen sense of geography and devilish courage could entertain few options for escape from this Confederate citadel that was surrounded by thousands of Southern troops and an elaborate system of fortifications reaching three miles south and west of the city. Some of the dozens of parapets were twenty feet thick, and some earthen redoubts were surrounded by ditches twenty feet deep and thirty feet wide. All of these were reasons to exercise discretion. But the absurdity of the harness incident and his humiliating confinement in the cellar pounding bricks into the mud floor provoked Wallace into one more desperate attempt at freedom. He fled the cellar and tells us in a passage full of retrospective pride, "Bid them all goodbye and went down into the city of Mobile."[44] Turnage was about to take impertinence to a new level, even for him.

For a week he "wandered . . . from one house to another where I had friends." This clandestine network in the black community, operating within and against the slave society that otherwise entrapped it, became Turnage's breathing room, his hideaway for the first week of his escape as he tried to plot his way out of the prison that was the Confederacy. As had John Washington, Turnage used the city to gain time to hide and hope, sleeping in horse barns or on the floor of willing black accomplices. But desperate teenage runaways could not rest anywhere for very long. One morning he was discovered in a stable by a "rebel policeman" with a pistol that was cocked and pressed against Turnage's breast. Dragged to jail by the neck and the seat of his pants in a "ridiculous manner," the fugitive was soon escorted to the "whipping house," and his master was called to the scene.[45]

Mobile's whipping house was the site of routine professional punishment meted out by hired hands who delivered lashes with whips "three leathers thick." The wartime Mobile newspapers were full of notices about the mayor handing out sentences of whipping to slaves who broke ordinances, violated curfew, committed petty theft, or stood accused of insubordination or inappropriate contact with whites. Slaveholders in turn were sometimes fined for allowing "too much liberty" to their slaves.[46] In the details of such sentences and the varying numbers of lashes prescribed, one finds not only the darkest corners of the world of slavery but also the vulnerable insecurity in the psyche of the master class. How much punishment could a slave owner demand and still retain his compliant laborer?

Turnage was stripped of his clothing as the elderly Minge watched. His feet and hands were tied with rope, and he was hoisted up on a hook set in the wall. First he was administered ten "licks" and allowed a brief respite as his body hung suspended. Then ten more were laid on by the man with the whip, who stopped again and asked Minge if that was enough. Minge answered: "Give him ten more."[47] Turnage calls this the "worst whipping I ever had." Released from the ropes, his skin sliced and bleeding, the slave was told to walk home.

Bloodied and tattered, but unbowed, Wallace had no intention of ever again serving Minge's family. He "took courage, prayed faithfully," and headed southwest toward an opening in the Confederate breastworks. Turnage hid briefly and then simply walked through an encampment of troops (who likely saw him as one of their slave servants or workers). Walking right through the opening in the trench works, the intrepid fugitive with the bloody back kept a good pace for what he describes as three miles across a terrain "cleared for a battle ground." Reaching a wooded area by dark, he spent the night without food, contemplating the trek ahead of him down the west side of the awesome Mobile Bay.[48]

In the most famous slave narrative of all, Frederick Douglass captures with masterful metaphors the anguish of the would-be runaway slave. As he peered north and dreamed of a route to freedom, Douglass saw a huge bay on his left, at once a frightful abyss and a watery hope. "The odds were fearful," Douglass declares. "At every gate through which we were to pass we saw a watchman—at every ferry a guard—on every bridge a sentinel—and in every wood a patrol. We were hemmed in upon every side." Turnage, too, contemplated a massive ocean bay on his left with the same conflicted sentiments. He had already seen more than his share of gates, guards, sentinels, and patrols, and now he was stepping through an army of them. Douglass's escape by land and water was the result of careful planning, good luck, and great courage; he later rendered its meaning with unforgettable prose poems. Turnage was living his prose poem as he stepped "barefooted" into the snake-infested swamps south of Mobile and gave its deepest meaning to this horrible war.[49]

Turnage's fifth and final escape came during the most fateful month of the war for Mobile, a time when the entire city "was in quite a stir." At the dawn of August 5, 1864, Admiral David Farragut sailed fourteen wooden ships and four ironclad monitors through a channel in Mobile Bay where the water was deep enough for such vessels. The channel ran close to the formidable Fort Morgan on the east edge of the bay, with its dozens of long-range guns. The mouth of the bay was mined with obstructions and torpedoes, one of which blew up the Union ironclad *Tecumseh*, sinking it in seconds and killing ninety-three Union sailors. But Farragut's ships surged on, pounding the huge Confederate ironclad *Tennessee* into submission in what turned into the largest and bloodiest naval battle of the war. While the city of Mobile would remain in Confederate hands until April 1865, the Confederacy's last major port was now closed, and the war-weary North had an important victory to celebrate.[50] Without the daring heroics of the Union navy in those dangerous waters and the army that stormed the forts, it is doubtful that Turnage's escape attempt later that month would have resulted in his freedom. In part, Wallace owed his own life to the 145 Union sailors who died securing the bay.

Turnage does not note the exact date of his final escape, but he twice indicates that his flight occurred in the "latter part" of the month of August. "I could see the warships of the Union lying way off in the pass and in Mobile Bay," he writes. Also crucial to Turnage's prospects were two other Union victories: the capture of Fort Powell, which was on a sand bar in the middle of Mississippi Sound, and of Fort Gaines, a brick-walled, well-armed structure on the eastern tip of Dauphin Island, the long barrier sand bar that guarded the southeast entrances to Mobile Bay. On August 3, General Gordon Granger had landed 2,400 Union infantry and numerous light artillery near the western extremity of Dauphin Island. By the morning of the great naval battle in the bay, Granger's men had trudged with their guns across several

miles of sand on the treeless island to within twelve hundred yards
of the fort and were poised to force its surrender. On August 5,
Granger, who would soon play a role in Turnage's fate, declared
his readiness for an assault on the fort, but only after honoring the
"severe" labor of his men in "a quagmire of deep heavy sand hot
enough during the day for roasting potatoes." Granger also reported taking into his command "some fifty splendid negroes"
who had been released by the Confederates in Fort Gaines. "Contrabands" would continue to flow out of the chaos and swamps into
Granger's lines and the Gulf Coast.[51]

On August 7, surrounded and under shelling from land and
sea, the Confederate commander of Fort Gaines, Colonel Charles
D. Anderson, agreed to surrender his 818 men without a fight, an
act for which he was severely condemned by his superiors. The
overall Confederate commander in Mobile, General Dabney H.
Maury, called the surrender of Fort Gaines "shameful." Moreover,
from far-off Richmond, President Jefferson Davis concluded, "The
surrender of Fort Gaines under the circumstances is deeply humiliating." And the famous Southern diarist Mary Chesnut reacted to
the news in equally gloomy terms: "Misery upon misery . . . Hor-
rid times ahead . . . Mobile half-taken."[52] Humiliation came in
many forms by 1864, and it was now a constant burden for most
Confederates.

Meanwhile, Turnage plunged on through the swamps on the
mainland above Dauphin Island, frightened and starving, but no
longer humiliated. Having passed by the large encampments of
Confederates, he still had to contend with patrols and picket
lines. As Turnage moved south, he would have either forded or
managed his way around the Dog River, probably in the area that
wartime maps listed as a "flat piney woods." He then crossed the

swampy Deer River. After four days without food and water, and
"troubled all day with . . . snakes," the wretched runaway reached
what for him was the fearsome Foul River estuary, which he
names and describes at some length.[53]

Today this extensive and beautiful wetland spawns the majority of marine life in the Gulf of Mexico. The landscape and wa-
terscape are both gentle and forbidding, with alligators crawling
through their wallows and laughing gulls squawking everywhere.
The delta grass—what Turnage calls "broom sage"—sways waist-
high in the hot summer breezes. After eating himself sick in a
cantaloupe patch and receiving one more charitable piece of food
from another wandering black man, Wallace "walked in the river."
As it deepened he swam as hard as he could and, by his account,
only narrowly survived the crossing. Feeling "very delighted," he
writes "that I had got across the river . . . I had dreaded so much."
He then hid in the delta grass and took stock of his next move.[54]

Turnage missed his opportunity for a proper baptism in Mobile, but he experienced a long baptism of another kind as he
inched closer to the ocean and the Yankee forces that he had been
seeking for three years. Rivers, with all their hopes and dangers,
are constant themes as both metaphor and descriptive reality in
many prewar slave narratives. Steep and slippery banks, currents
that overwhelm human strength, and the sighting of distant shores
draw readers into the plight of the runaway slave. "I entered the
river . . . plunged into the current," remembers Charles Ball of his
journey out of North Carolina. Ball's river was a "violent . . . mad
stream" full of ice in the dead of winter. The current that nearly
destroyed him then saved him as he grasped some branches and
"crawled" ashore. It is likely that former slaves such as Wallace
Turnage and John Washington had heard the story of the young

Jesus of Nazareth, who, after being baptized in the River Jordan by John, "saw the heavens opened, and the spirit like a dove descending upon him." That spirit renewed Jesus and drove him into forty purposeful days in the wilderness. Rivers provided this spirit of baptism and renewal for slaves, especially once they were on the other side. Wallace might not last another forty days in his wilderness, but he had not given up. And he insists that "angels" came to save him from "Satan" and "wild beasts," just as they did for Jesus in the Gospels.[55]

Hoping he was now safely across his river, Turnage spotted a rifleman on horseback. In remarkable detail, he describes the "shine" from the sun on the horseman's gun that caught his eye and forced him to "crawl" once again into a salt marsh, where he hid for another day suspended just above the swamp water on bushes he had bent to make himself a bed.[56] Whether wading, crawling, swimming, or desperately scrambling in the muck, Turnage now knew where he was going. He wanted to see the gleam of Yankee guns.

When Turnage finally reached the southern tip of the mainland at Cedar Point, he had traveled a full twenty-five miles south of Mobile. Here he could look out to the sea and see Fort Powell. Making a "hiding place in the ditch of the old fortifications," he ducked Confederate sentinels who still occuped the area, continued to starve for another full week, and grew "so impatient seeing the free country in view and I still in the slave country." In the last pages of his narrative, Turnage becomes a mystical character in the hands of the divine, as well as a memoirist who converts his story into one stark, agonizing contrast between slavery and freedom after another. The sun may have been blinding and his body all but spent, but Wallace's choices were clear: "It was death to go back and it was death to stay there and freedom was before me; it could only be death to go forward if I was caught and freedom if I escaped."[57] This was Turnage's own timeless expression of the human will to choose freedom over death and slavery found in most slave narratives. He would keep the hero's character uncomplicated and his courage pure while the ocean itself provided the drama.

But Turnage was a desperate hero. He barely survived his first attempt at a crossing; he tried to ride a log into the ocean and only narrowly made it back to shore when his pole proved too short to reach the bottom. Hiding again in his "den," he now played a game with the Confederate soldiers who came by day to occupy what Turnage called their "high spie house," perhaps a platform built up in a tree. The runaway slept in their lookout by night to avoid the mosquitoes and climbed down in the mornings before his enemies arrived for duty. Turnage even uses remembered sounds as he struggles to convert his memories into prose. Back asleep in his ditch one morning, "the clinking of a rebel sabre climbing up into that spie house" awakens him and reminds him of God's protective hand.[58]

Almost as though remembering a verse from a spiritual or anticipating a future blues song, Turnage describes himself as "very much troubled in mind" as he confronted death while hiding on that barren shore. So he "prayed for deliverance," and before the next morning he heard a "voice singing within me." He descended from the lookout, walked to the water's edge, and saw there, as if "held by an invisible hand," a "little boat" that had rolled in with the tide.[59]

The veteran runaway found a "piece of board" to use as an oar and began to row toward Fort Powell, bailing out his boat as he went. A "squall" came in sight, "the water like a hill coming with a white cap on it." As the heavy sea struck his boat, Turnage lost control and was nearly "swamped"; then suddenly he "heard the crash of oars and behold there was eight Yankees in a boat." This time sound and sight meant exhilaration rather than fear. Keeping rhythm with the movement of the oars, the spry Turnage jumped into the Union gunboat as his own little vessel "turned bottom up" in the crashing waves. For a stunning few moments the oarsmen in blue "were struck with silence" as they contemplated the frail young slave crouched in front of them. Wallace turned and looked back at the Cedar Point shore, where he could see the Confederates peering out at him, and measured the distance of his bravery at sea. Then, as the liberators' boat bounced on the waves, he took his first breaths of freedom.[60]

Half an hour later Turnage was safely in the ruins of Fort Powell with the Union garrison. The troops asked him to tell his story, invited him into their tents, gave him food, and extended him "attention in all that I wanted." That night he slept in a Yankee camp, the long-imagined destination of his dreams. The following day, likely August 24, 1864, Turnage was taken in a skiff to Fort Gaines on Dauphin Island and, remarkably, brought before General Gordon Granger, who was fresh from the triumphant capture of Fort Morgan after a ten-day siege.[61]

Granger was a forty-four-year-old West Pointer and veteran of the Mexican War. Strikingly handsome, Granger was known to be independent-minded and skilled in combat. He was a keen supporter of emancipation and the use of black soldiers in the Union army, a policy which was by now a full year in operation. That very month he had requested the deployment of the Ninety-seventh and Ninety-ninth United States Colored Troops in his command. Granger probably interviewed Turnage because Wallace had journeyed all the way from Mobile and would have had knowledge of the Confederate forces and the civilian conditions in the city. Either in a bunker office of the brick fort or out in its interior courtyard, Granger gave Turnage two choices: He could either enlist in the army or hire himself out as a servant to a Union officer. The decision seems to have been easy for the scarred fugitive. His bravado aside, Turnage had seen enough of his own private war against Confederate troops, bloodhounds, overseers' whips and chains, snake-infested rivers, and ocean waves. He opted to serve as a cook for a Marylander, Lieutenant Junius Thomas Turner.[62]

A native of Baltimore, Turner may have enlisted as a drummer boy during the Mexican War. He migrated to California during the Gold Rush and attended the University of the Pacific in Santa Clara in 1851 and 1852. Turner pursued various professions—lawyer, accountant, probate judge—until he enlisted as a private in the Fourth California Infantry in 1862. Discharged with tuberculosis, he reenlisted a year later in the Second Massachusetts Cavalry, a regiment comprised of East Coast natives who paid their own way back East and served in northern Virginia. Discharged again after a horse kicked him and broke his leg, he convalesced and took a commission in the Third Maryland Cavalry, which he joined in Thibodaux, Louisiana, in May 1864. Three months later Turner met Wallace Turnage on Dauphin Island. As the war ended in the spring of 1865, the two were delivering Confederate prisoners of war to New Orleans. Promoted to captain, Turner served in many other capacities during his summer in the Mississippi Valley; and Turnage probably stayed with him.[63]

Frustratingly, after recreating the drama of his final escape, Turnage quickly brings his narrative to a close by telling us only that he accompanied Turner through the capture of Mobile on April 12, 1865, and then on to New Orleans and up the Mississippi River to Vicksburg and Natchez. He and Turner may have parted ways in Mississippi in August when the captain returned east due to bad health, but there is some evidence that Wallace traveled with the Maryland regiment all across the country and was present for its mustering out in Baltimore on September 7, 1865.[64]

As a free man on his own, Turnage reunited with his North Carolina kinfolk and strove to make a new life. He does not tell us what he thought of his liberty in that first summer after the war and emancipation, but we might draw our own conclusions from his statement at the end of the narrative. As though answering the question of an interviewer, he leaves this stunning statement about natural rights and the meaning of freedom: "I had made my escape with safety after such a long struggle and had obtained that freedom which I desired so long. I now dreaded the gun and handcuffs and pistols no more. Nor the blewing [sic] of horns and the running of hounds; nor the threats of death from the rebels' authority. I could now speak my opinion to men of all grades and colors, and no one to question my right to speak." As one of the "many thousands gone," Wallace Turnage, in a form of folk poetry, crafted his own emancipation hymn.[65]

CHAPTER THREE

Unusual Evidence

> If the Negro had been silent, his very presence would
> have announced his plight. He was not silent. He was in
> unusual evidence. He was writing petitions, making
> speeches, parading with returned soldiers, reciting his
> adventures as slave and freeman.
> —W. E. B. DuBois, *Black Reconstruction in America*, 1935

Throughout African American history—by force and by choice—migration has been a constant reality. Across oceans, over rivers, in wagons loaded down with worldly possessions, on trains, by foot over forbidding landscapes of war and wilderness, and from country to city, movement to new lands and new places has shaped the story of millions. By 1862 and 1865, respectively, John Washington and Wallace Turnage were immigrants, boldly participating in the flow of emancipation—what W. E. B. DuBois called "this slow, stubborn mutiny of the Negro slave."[1] Their mutiny completed, Washington and Turnage now faced the daunting task of forging lives for themselves and their families at the bottom rungs of the emerging urban black working class in Civil War America.

Both former slaves would have remained unknown names on census rolls—their migration and resettlement stories lost to time—had they not recorded their emancipation narratives. They would have remained anonymous laborers, members of the working poor who strove to raise families, garner respectability, educate their children, and save some meager earnings, hoping that their offspring might somehow break the bonds of class and race in Northern cities. The overwhelming majority of the freedpeople remained in the South after the Civil War and emancipation; approximately 80 percent of those who stayed in the South became sharecroppers by the 1870s and 1880s, with largely tragic results.

Washington and Turnage, though, were among the minority who, by the force of war as well as their own wits, moved to the North. So far as can be determined, neither man wrote about his postwar life. But the existence of their narratives and the fierce desire to write them in the face of their own literary and educational limitations surely demonstrates their will to be known. "If the Negro had been silent," wrote DuBois in 1935, "his very presence would have announced his plight. He was not silent. He was in *unusual evidence*. [Italics added.] He was writing petitions, making speeches, parading with returned soldiers, reciting his adventures as slave and freeman."[2] Washington and Turnage were almost silenced forever by time. But they have now become the *unusual evidence* that DuBois imagined. They can be retrieved from obscurity by fragments of data—a bank record here, a city directory there, brief obituaries, pension applications, a cemetery headstone, and some precious photographs.

We know where Washington and Turnage lived, but can confirm only frustratingly small pieces of how they lived. We don't know the precise rhythms of their work lives or the specific dramas of their growing and struggling families. But we do know their various occupations, many of their addresses, their church and fraternal-order memberships, and we have some vestiges of their civic lives. We know Washington died in 1918 and is buried in Woodside Cemetery in Cohasset, Massachusetts; and we know Turnage died in 1916 and is buried in Cypress Hills Cemetery in Brooklyn, New York. We can illuminate some small parts of their long postwar lives even though large elements remain hidden. We can be sure that the central daily dilemma of their lives was finding meaningful gainful work, especially in the early years of their freedom. Washington and Turnage exemplified DuBois's claim in 1901 that most African Americans had "emerged from slavery into a serfdom of poverty and restricted rights."[3] But we can also see them in photographs, wearing their best clothes, demonstrating their freedom and respectability, transcending the racial strictures of the Gilded Age and Jim Crow America. And many of their children carried on the legacies of striving and independence learned from their formerly enslaved fathers.

Washington and his extended family arrived in the District of Columbia during the desperate Union disaster at Second Manassas. In the immediate wake of this bloody battle, the Union commander, Major General John Pope, was near a mental breakdown. He declared his troops "badly demoralized" and feared they would "melt away" as they scrambled to retreat within the capital's fortifications. In desperation, President Lincoln restored the sanctimonious General McClellan to overall command of the Union armies defending Washington. McClellan hardly could have grasped the irony in his dispatch of September 1 when he urged his officers to help Pope's army. "This week is the crisis of our fate," declared the pro-slavery general, who openly opposed emancipation in any form.[4] All over the region waves of freedmen were advancing or retreating just like the soldiers, fully aware that these weeks were the crisis of their fate as well.

Meanwhile the city swelled with wounded men and escaped slaves. Crews worked for days to bury the more than three thousand dead lying on the ghastly battlefields around Manassas. Wounded soldiers occupied two thousand cots in the U.S. Capitol building, as well as in churches, a college, and a seminary that were commandeered as hospitals. Another thousand Union wounded remained in field hospitals in Alexandria, Virginia, just across the Potomac River from Washington, where a growing "contraband camp" full of slave refugees had emerged in terrible conditions. Other camps sprung up around the city; freedpeople also set up tar-paper shanties and took up residence in barracks previously inhabited by troops. One such site was Camp Barker, an old barracks and occasional prison converted into a tent city for former slaves from rural Maryland and Virginia, located at Twelfth and Q Streets, near the current-day Logan Circle. Another contraband camp, known as Freedmen's Village, grew under the supervision of civilian-funded freedmen's aid societies near the mansion of the Confederacy's commanding general, Robert E. Lee, just across the Potomac River in Arlington (the future site of Arlington National Cemetery). Eventually, the roughly fifteen hundred ex-slaves lived in barracklike dwellings in the Village. By early 1863 approximately ten thousand freedmen had made Washington, D.C., their new home, and by war's end that number had escalated to an estimated forty thousand.[5]

In this environment, John Washington sought work and fought to survive while he and his family became part of the "first freed," as many in the growing African American community came to call themselves. Some blacks were employed temporarily in the city's burgeoning hospitals, and it is likely that Washington worked at least briefly in support of the wounded.

John and Annie were part of the community of some four hundred members, most of them former slaves, of the Shiloh Baptist Church in Fredericksburg who had fled to Washington in 1862 and 1863. They had been married in that church on the Rappahannock, and they were likely part of the first gathering in a "little shanty" when the congregation reestablished itself on L Street, between Sixteenth and Seventeeth streets just four blocks north of the White House. Shiloh was formally recognized by other Baptists on September 27, 1863, the Sunday during the week of the first anniversary of the Preliminary Emancipation Proclamation, a date Washington and his fellow parishioners surely would have celebrated. Under the leadership of Reverend William J. Walker (himself from Fredericksburg and whom Washington mentions in his narrative), the church grew to eight hundred members, purchased two buildings on L Street, and eventually constructed a two-story brick edifice in 1883. For at least the first two decades of Washington's residence in the District, Shiloh Church was the center of his social world. Annie sometimes sold wares at the church's many fairs, some of which lasted for weeks at a time. If they ever could have left their young sons in someone's care they might have joined one of the church's occasional excursions back to old hometowns in Virginia, such as the one to Fredericksburg and Richmond in August 1885. And the Washington children did not likely miss many of the church picnics.[6]

During the early 1870s John served the church as clerk, as well as superintendent of Sunday schools. According to one account, he directed the Sunday schools for twelve years when the attendees were mainly black adults, many of them former slaves, and the teachers were both black and white. Washington was also active in the Baptist Sunday School Union, which an 1885 *Washington Bee* article described as "the strongest religious organization in the District of Columbia." The Union engaged in a wide array of civic work for the emerging black community, including establishing a normal school (primary education), creating a library, and starting a journal, as well as raising money to provide shoes and clothing for poor children. In 1879 Washington served as the Union's vice president, and he chaired a committee that conducted a statistical study of all Baptist Sunday schools in the District. And by 1900 it appears that John and Annie were members of an order of black Masons and its women's auxiliary.[7] Clearly, John put his literacy, his faith in education, his civic spirit, and his work ethic to the service of his new community. Employed by day as a common laborer and eventually a house painter, by night he became through the force of his will a family man and citizen John Washington.

By the late 1880s the Washingtons had seceded from Shiloh Baptist along with other disgruntled parishioners. John and Annie moved to a house on I Street and joined the Nineteenth Street Baptist Church, directly across the street.[8] Whichever Baptist congregation they embraced, and whichever minister they followed, the Washingtons made the church their spiritual and social center. In this one independent institution that they had brought with them from slavery, freedpeople educated children, nurtured a community increasingly surrounded by a racially hostile world, found spiritual solace, gathered their scattered families, and took ownership of their lives. In their churches the Washingtons, like so many other families of freedmen, negotiated the private and public dimensions of a racially obsessed society. Inside God's house they were utterly free; outside they were strengthened for the struggle against Jim Crow's cunning ways.

Washington first appears in a city directory in 1864, his occupation listed as waiter. He lived at 311 Eighteenth Street, now the site of Constitution Hall and a mere two blocks southwest of the White House, or Executive Mansion as it was then called. And he watched intently as three regiments of the United States Colored Troops were organized in the District of Columbia in 1863 and 1864. Washington might have attended some of the numerous recruitment meetings and heard the rousing rhetoric of such orators as George W. Hatton, a former slave and clerk in a pharmacy, who became a corporal in the First USCT. Wounded in the knee in battle, Hatton became a local hero and a city councilman in the District.[9] There are several explanations for why Washington did not join one of these regiments and march back into Virginia to fight: He was small in stature, had a young family to support, and had already seen a good deal of war. In "Memorys of the Past," he betrays no special desire for martial glory, and he had already given nearly four months of service to General King's staff.

On April 16, 1863, the first of what would become a long and storied tradition of D.C. Emancipation Day celebrations was held. Each year on the anniversary, blacks in ever larger numbers held parades, conducted rallies with speeches, celebrated themselves through their civic and fraternal organizations, dressed to the nines, gave the city a high-stepping musical spectacle, and flexed their political muscle. The First Ward Republican Club, which held its meetings at Washington's church, Nineteenth Street Baptist, organized these parades every April from 1866 into the 1870s.

John would have been among the fifteen thousand people at the 1866 celebration, and he may have helped in creating and painting some of the dozens of banners honoring LINCOLN THE EMANCIPA-TOR or those that read UNION AND LIBERTY FOREVER.[10]

The vibrancy of this new black community was on full display at these parades, as was its politics. In the face of stiff and some-times violent opposition from whites, District blacks demanded and exercised their right to vote. In late 1865, while the all-white Washington city council announced it their "Christian duty" to oppose black suffrage in any form, the black community secured 2,500 signatures on a petition they sent to Congress insisting on passage of a suffrage bill. Since the organizers wanted literate folks to sign the petition in order to show the respectability of the com-munity, John was likely a signer. As Congress debated the great questions of Reconstruction policy in the spring of 1866, blacks in Washington faced a hostile white press that condemned their community for its "vagabond habits . . . theft, burglary, arson, and similar crimes." "Should Congress pass the suffrage bill," warned the *Evening Star,* "the District promises to be the Botony Bay of idle, vicious, darkey-dom." The June 1867 municipal elections were the first in which black men voted, and the surge of registration was astounding. That year, 9,792 whites registered to vote in the District, but so did 8,212 blacks—and all of the latter were Repub-licans. On the morning of the elections, groups called out voters with horns and bugles in black neighborhoods; at one polling place in John's neighborhood the line was eight hundred strong before any whites had arrived. In the 1868 election, blacks carried the First Ward, where Washington lived, and elected John F. Cook as alder-man and Carter A. Stewart as councilman.[11] Such political activ-ity and public declaration of their rights as citizens energized young freedmen like Washington to rise each morning, face their poverty, and sustain the hope that their children would live to see a better day.

Washington's black community divided over the nature of the Emancipation Day parades in the 1880s. Middle-class folk came to see them as too rowdy and bawdy, and the tradition eventually waned and nearly died—but not before nearly two decades of extraordinary public expressions of the meaning of freedom and liberty. Spectacular floats were the order of the day in the 1860s and 1870s, as were hundreds of banners with competing slogans and remarkable artwork. John Washington may have been among the group of First Ward Union Leaguers in 1867 that carried an especially fine white silk banner emblazoned with a large portrait of Lincoln. Or he may have simply joined the crowds lining the parade routes, holding the hands of his sons, watching the militia units and marching bands prance and the chariots rolling by. He also might have pushed his way into the packed Fifteenth Street Presbyterian Church in 1883 when District resident Frederick Douglass delivered a masterful speech about the meaning and memory of emancipation. Or perhaps John attended Douglass's stunning condemnation of the nation's betrayal of black life and liberty in 1888. At that evening's commemoration, Douglass accused the federal government and even his own party, the Republicans, of treating the freedman as "a deserted, a defrauded, a swindled outcast; in law, free; in fact, a slave; in law, a citizen; in fact, an alien; in law, a voter; in fact, a disfranchised man."[12]

The new day for which all the John Washingtons yearned was very much in jeopardy as the white South regained control of their society and politics and crushed the rights of blacks by intimidation, murder, and disfranchisement. In John Washington's neighborhood, his children could still attend school and hope for a life of mobility and dignity, but only under the ever-darkening cloud of American racism, and in a culture increasingly forgetful of the deepest causes and consequences of the Civil War. Freedmen such as Washington would always know what Douglass so frequently proclaimed in the waning years of the nineteenth century—that "slavery has always had a better memory than freedom, and was always a better hater."[13] Perhaps John wrote his narrative in part because of his own personal sense of this discordance between his memory and that of the nation in which he now claimed such a tenuous hold on citizenship.

Washington's greatest preoccupation, however, was with the daily struggle to reassemble his family. Sometime in 1865, in what must have been a joyous reunion, John's mother and her husband, Thomas Tucker, arrived in Washington from Staunton, Virginia, and took up residence in Georgetown. John and his mother had not seen each other in fifteen years. Sarah had finally achieved freedom by the end of the war, perhaps in part by virtue of the capture of her owner-employer, the Reverend Richard Phillips, who had joined the Confederate army and had been imprisoned in 1864 at both Camp Chase in Ohio and Point Lookout in Maryland. Exactly how and when Sarah and her son communicated and found one another is left to our imagination. They probably wrote letters, some of which must have made it through the war lines. In addition to asking friends for news, many former slaves advertised in black newspapers for information about loved ones, and it is possible that John used this means of contacting his mother as well.

Washington was lucky; this reunion of mother and son was among thousands realized all over the American landscape of emancipation, but thousands more went unrealized. One has to wonder if John took his mother for a stroll around the White House grounds and tearfully recollected the moment when she lay with him on the bed at the Farmer's Bank in 1850 before she departed with his four younger siblings. They may have found the demands of daily life too burdensome for this kind of reflection, but the two former runaways comprised an intergenerational storehouse of both the memory of slavery and the meaning of freedom. Whatever hardships their new lives in Washington brought them, John and his mother could look up from the dirty streets of the city and yearn for a good Virginia ham or some other mystic remembrance of their roots; but they also could gaze up at the edifices of the nation's capital, read and hear about the great debates in Congress over Reconstruction, and believe that the promise of the new nation belonged to them even though they were poor and carrying a past of enslavement. When the Fourteenth Amendment was ratified in 1869, enshrining "equal protection of the law" into the U.S. Constitution, John Washington could have gripped Sarah's hand on one side and Annie's on the other, and uttered a collective hallelujah. And the next morning two generations of new citizens through whom a "second founding" of the American republic took form would go back to work.[14]

Tragedy and loss also struck Washington's family in these early years of freedom. Sometime between emancipation in 1862 and the early 1870s, grandmother Molly died.[15] That his grandmother could live out her last days in freedom, however harsh the life, must have been a source of pride for John. But the Washingtons suffered an even greater loss in March 1865, only weeks before the end of the war, when their second child, John Burnside Washington, died three months before his first birthday.

In May, when spring flowers were available for the child's grave, Washington wrote a four-page eulogy—a mourning prayer— about the death of his son. Entitled "The Death of Our Little Johnnie," this rare document survives along with his narrative of emancipation (see appendix). "Who has not lost some dear little one who used to be the joy of the household," writes the grieving father. The eulogy, like the narrative, demonstrates Washington's deep need to write sorrow and joy into and out of his soul. Alternating between quoting scripture and describing the child's demise, Washington gives voice to his pain: "For Christ says 'Suffer little children to come unto me and forbid them not, for of such is the Kingdom of Heaven.'" Then he observes the child on the death bed in the typically nineteenth-century way: "After being racked with pains and wasteing fevers until the little voice is scarcely audible, the dear little bright eyes is closed never more to look up to see; his hands sag by his side motionless and he cannot talk to tell you of his necessities." Writing as a parent, John found a poetic voice and a redemptive mood. "When you can only think for the little sufferer and view his agony without any power to help," he laments, "what a condition for the parents to be placed in." The scale of wartime death had deepened Washington's religious faith, which helped him find meaning even in the loss of such an innocent. "The messenger of death comes and relieves the little sufferer from all earthly pain," writes the father, "and the Battle just begun is won, and the Victory complete. And white winged angels waft its spirit upward to the skies, where there is no sin and there is no weeping there; 'for Christ has said of such is the Kingdom of Heaven.'"[16]

In a nation overwhelmed by mourning, the twenty-seven-year-old John Washington, a sorrowful father and former slave with a hard journey behind him, desperately needed to at least face, if not understand, the death of his free-born infant son. At Johnnie's grave he acknowledged the victory over the "bleak winds" and "mournful dirges" of winter by the "sweet perfumes of spring flowers" and the "little spires of green grass" pointing "upward to God." All over America, millions yearned for such faith and consolation in the face of overwhelming loss. Over 600,000 lay dead from the war, most in unmarked graves, and more than a million were maimed and incapacitated. The assassinated Abraham Lincoln was laid to rest in Springfield, Illinois, six days before Washington wrote his death prayer. As an avid reader of newspapers, Washington surely followed the story of the president's funeral train as it wound its way across the Northern states. The freedman father's lament cannot help but remind us of Walt Whitman's remembrance of Lincoln in "When Lilacs Last in the Dooryard Bloom'd." Flowers were blooming in the cemetery in Georgetown where John and Annie buried their tiny son. They felt the same sweet shuddering that Whitman felt for the nation and all those who grieve and hope for springs to conquer winters:

> With many a pointed blossom rising delicate, with the
> perfume strong I love,
> With every leaf a miracle—and from this bush in the
> dooryard,
> With delicate-color'd blossoms and heart-shaped leaves of
> rich green,
> A sprig with its flower I break.

Washington could not match Whitman's "powerful psalm in the night," but he mustered from his son's death a faith in a better place where, he imagined, "the weary ceased trembling."[17] Death can level all distinctions; the president and little Johnnie, after all, had been neighbors.

John and Annie continued to raise their family in the Foggy Bottom section of Washington, the ten-block area west of the White House where George Washington University stands today. On January 12, 1866, only a few weeks after Congress ratified the Thirteenth Amendment abolishing slavery and convened the Joint Committee on Reconstruction to assess the fate of the seceded states and the freedmen, James Arthur Washington was born to Annie and John, who was laboring as a "barkeeper."[18]

In 1867 Washington opened his first account at the Freedmen's Savings Bank, listing himself as secretary of the Daughters of Shiloh Church, among other things. That year, and for at least the next thirty-eight years, his occupation appeared in the city directories of the District as "painter." Undoubtedly Washington performed all manner of interior and exterior painting, but in his 1918 obituary he would be listed as a "retired sign painter," implying that he did fine commercial work as well. In 1868 he and Annie moved to K Street near Twentieth Street, and a third son, John M. Washington, Jr., was born. A year later they had moved one block west to Twenty-first Street, and Annie appeared in the directory as a "dressmaker."[19] Two incomes were essential for this emergent working-class black family, especially if they aspired to education and social mobility for their children.

Typical of the economic insecurity and high birthrate of the new urban working class in America, the Washingtons would move at least four more times in the coming decades and give birth to two more sons, Charles Somerville, born November 9, 1870, and Benjamin, born June 13, 1873. After languishing in bed for at least six months, Sarah died in 1880 at the age of sixty-two and was laid to rest in Mount Zion Cemetery in Georgetown.[20] John had now buried his second-born son, his grandmother, and his mother in the free soil of the nation's capital.

For freedpeople, the legacies of slavery and the new births of freedom implied by the Constitutional revolutions of early Reconstruction meant so much more than the abstractions through which we often understand that experience today. For them it was a world full of daily struggle and hope, of ends and beginnings, of work and patience. The hard-working painter and his busy wife, raising five sons ranging in age from newborn to eleven, had little time for historical reflection in 1873. But that, indeed, was the year Washington sat down to write his narrative. Certainly one of his most important reasons for writing was to leave his story to that brood of sons who otherwise might never really know where they came from. Using his pen strokes instead of the brush strokes of an itinerant painter, John Washington wanted his sons to understand that he was much more than the common laborer they saw working in the alleys of Foggy Bottom.

As the Washingtons' sons grew to adulthood and strove for respectability, they were privileged as light-skinned freedmen's sons by virtue of their living in a city that, despite racism and discrimination, forged a solid black community with a faith in education. We can only wonder how closely the sons read their father's narrative. Did they embrace it with pride or shun it to avoid the stigma of slavery? Did such striving allow them to fully appreciate their parents' generation as the founders of a new day and a new history in African American life?

Most of the sons moved away, seeking all the geographical and class mobility that was possible in this darkening age of race relations. None, of course, moved southward into the snares of Jim Crow and the land of lynching, although the North was hardly devoid of either by the 1890s. Early that decade Charles moved to Chicago, where he worked as a janitor and a bank messenger at the Corn Exchange Bank before owning a lunch café in the Loop with his wife, Mabel. In the same decade, John Jr. relocated to New York City and worked as a barber and a railroad waiter. After living with his parents until 1900, William, the oldest son, moved to Boston, where he became a tailor and a shipping clerk, and married his neighbor, Louisa Lewis. At some point James, a railroad worker, moved with his wife, Catherine, to Cohasset, Massachusetts, an oceanside town south of Boston. Some of the Washington sons remained in jobs assigned to black workers (barbering, railroad workers, waiters) while others achieved professional status. William relocated with his wife to Chicago sometime after 1910 and opened a real-estate office in the midst of the Great Migration to that city during and after World War I.[21]

From 1911 to the 1930s Charles and Mabel, as well as William and Louisa, appeared frequently in the society pages of the *Chicago Defender,* a leading black newspaper. Charles and Mabel lived in a "beautiful home" on Bowen Street and later moved to Calumet Street, hosting or participating in all manner of dinners, parties, charities, and dance clubs. Mabel was a prominent club woman and often hosted a women's group, The Classique Five Hundred Club. Even their occasional vacations to southwest Michigan and Annie's visit in 1912 were noted in the *Defender* columns as examples of the steady, accomplished class among blacks. During the nadir of the black experience—the Jim Crow era—the small prospective black middle class constantly trumpeted the "progress of the race," measuring success for their own reassurance and in the hopes that white Americans would notice.[22]

This was a fluid and dangerous time in race relations. John and Annie Washington taught their sons ambition, and the young men went out to learn for themselves the truth Ralph Ellison would later describe: "For the slaves had learned through the repetition of group experience that freedom was to be attained through geographical movement, and that freedom required one to risk his life against the unknown."[23] The Washingtons' sons understood the depth of their generational heritage from their slave father and, like him, became migrants looking for new chances and new homes.

The youngest son, Benjamin, achieved an advanced education and a truly professional life. He graduated from the prestigious M Street High School in 1895 and the Minor Normal School in 1896. Most of the children of the District of Columbia's black elite attended M Street High School. It eventually boasted of many famous graduates, including Robert H. Terrell, Roscoe Conkling Bruce, Benjamin O. Davis, Rayford Logan, and Charles Hamilton Houston. The school also attracted a talented array of black teachers, including Carter Woodson, Mary Church (Terrell), and Anna J. Cooper. Some of its graduates went on to attend America's most elite colleges, such as Harvard, Amherst, Oberlin, Princeton, and Dartmouth. The District's public school system, although segregated, was unique in that the schools for black students developed independently and had largely black leadership, with nearly equal salaries for black and white teachers. Nonetheless, it was segregated, overcrowded, and fraught with internecine controversies over appointments of teachers and principals. M Street had outgrown its capacity of 444 students by 1895, the year Benjamin Washington graduated. And its thirty-one full-time and three part-time teachers had only twenty-one classrooms and laboratories for 618 students.[24]

To the aspiring black middle class of Washington, D.C., the public schools were of utmost importance. The young people privileged enough to get into M Street High School were often the sons and daughters of "race men" and "race women," and they were considered the future leaders in the parallel "colored" universe of Jim Crow America. John and Annie Washington were hardly members of the District's famed "black aristocrats," but that their youngest son made it to M Street and beyond is a testament to their values and aspirations.

Benjamin appears to have taken the cause of uplift to heart. At various times he attended summer courses at the Armour Institute of Technology in Chicago, the Massachusetts Institute of Technology, Harvard University, and the Hampton Institute in Virginia. In 1903 he received a Bachelor of Pedagogy degree from the Teachers' College at Howard University in Washington and then took a job at the Armstrong Manual Training School (later Armstrong Technical High School), where for the next forty years he taught general science, physics, and history. In 1893, while in high school, Benjamin also was a proud charter member of the Washington High School Cadets, a marching drill team that for the next fifty years he would help sponsor. In its first competition, he commanded the winning company and prepared the commemorative list of all participants.[25] Benjamin seems to have inherited his father's organizational acumen and community engagement.

In his varied civic career, Benjamin helped found the Twelfth Street YMCA and served on its board for twenty-five years. Moreover, he coached numerous sports at Armstrong Technical, and from the 1910s to 1950s he served as commissioner of officials of the Central Intercollegiate Athletic Association, a group of twenty-six Eastern black colleges. Along with the soon-to-be-famous sportswriter Sam Lacy, Benjamin lodged a protest in 1927 when he and three other blacks were suddenly dropped from the rolls of the American Athletic Union's list of approved basketball officials. At that juncture Benjamin was referred to in a press report as the "father of local basketball" in Washington, D.C., as well as the man responsible for instituting the "free throw" after personal fouls. In 1955, the *Washington Post* reported that not only had the CIAA dedicated its annual meeting to Benjamin as he stepped down from leadership due to ill health at age eighty-three, but that the eighteen-team football conference had marked a "turning point in its history" by considering resolutions to face the "effects of integration" following the landmark *Brown v. Board of Education* Supreme Court decision of the previous year. The athletic conference now allowed its schools to participate in both black and integrated conferences at the same time.[26] Today we might note that such a person—a science teacher for four decades and a commissioner of a major athletic conference—was the son of an American slave. In the 1950s, however, when the remembrance and study of slavery had not yet undergone the revolution it would experience in the next three decades, Benjamin's paternity and historical roots went unacknowledged.

By any standard, John and Annie's youngest would have made them proud. His mother had been born free, and it is possible that the Washington sons identified as much with her as with the father who had left them his emancipation narrative. In 1907, Benjamin bought a house at 936 S Street in Washington, and his aging parents moved in with him. Their next-door neighbor was a black man, Daniel A. P. Murray, an assistant librarian of Congress and very active in District civic affairs. The next year Benjamin married Mary E. (Mamie) Grimshaw, whose family members also worked at the Library of Congress. By 1912, the fiftieth year since John Washington's emancipation, Benjamin and Mamie had their only child, Evelyn. It was this granddaughter, Evelyn Washington Easterly, who would end up in possession of John's narrative in the 1970s and who would pass it on to her close friend Alice Jackson Stuart.[27] The family heirloom may have been one precious way a granddaughter could connect with the grandfather she hardly could have known.

Benjamin also carried on the family's church traditions, serving as a member and vice chairman of the board of trustees of Nineteenth Street Baptist Church in Washington, as well as one of its choir directors from 1925 to 1939. Among the Washingtons' fellow congregants at Nineteenth Street Baptist in the first decade of the twentieth century was the youthful Edward "Duke" Ellington, who experienced some of his first musical influences in that church.[28]

In biography we often find our best subjects not only among the famous and powerful, but also among those whose lives are laced with colorful drama, personal turbulence, heroic triumphs, and desperate failures. In the case of John Washington and his sons, the story seems more one of striving for respectability than for notoriety—for sturdiness and security, for paths of virtue and community and family uplift for their own sake. For families of freedpeople bred out of the violent rebirth of emancipation, sturdiness behind the nation's segregated veil was its own heroic triumph. Always on some level an advocate of manual training and self-help among blacks, Benjamin nevertheless was a member of the National Association for the Advancement of Colored People, an organization devoted to protest. He died at Freedmen's Hospital in Washington on December 9, 1957, the year the Little Rock Nine integrated Central High School by bravely marching before white mobs and television cameras.[29] Links between the first and second emancipations in America are closer than we sometimes imagine.

In 1913 John and Annie Washington considered themselves retired and moved north to live with their son James and his wife, Catherine, in Cohasset, Massachusetts. In their old age, John and Annie likely needed someone's care, and Benjamin had just started a family in a small house. So the move to Massachusetts from bustling Washington must have been both welcome and a cause of great ambivalence. They had never lived anywhere else since 1862. James had moved to the small, thriving seaside town— already a resort and summer playground for rich Bostonians— sometime in the first decade of the twentieth century. It is extraordinary that a young black man would move to such a town, but he found a secure job working for the Old Colony Railroad, an extension of the New York, New Haven, and Hartford Railroad. James purchased a stunning three-story house at 312 North Main Street that was set on a hill some sixty yards from the road, on three acres. The scene would have felt rather rural in 1913, as it does today. The house was less than a mile from the center of the village of Cohasset and even closer to the Black Rock railroad station at the intersection of King and Main streets, where James worked for more than forty years as the crossing attendant, operating a hand crank that raised and lowered a traffic barrier and kept the trains running safely.[30]

It was in this house, with a nice view and a wraparound porch, that John Washington lived his final five years. John and Annie witnessed parades, fairs, and flag raisings in the village during World War I. And perhaps they occasionally enjoyed carriage rides out on Jerusalem Road to the seaside to see the rows of summer "cottages"—mansions built between the 1880s and 1910s by leather barons and other wealthy Bostonians. They probably remained less conspicuous than one of the few other black residents of Cohasset, Tom Loney, a former slave from Virginia, who drove his "Jerusalem Road Swill Cart" to collect the garbage of rich folks in the big houses. They no doubt became well acquainted with Thomas and Carrie Bowser, the only other prominent black family in Cohasset, who lived just two addresses to the south of them on North Main. We do not know how many of the permanent residents of the village—many of whom were Irish, Italian, and Portuguese Azoreans—and the well-heeled blue bloods taking the summer sea air knew that this former slave and Union army servant lived among them. Some among the Italian immigrant Garofolo family, however, must have later learned John's story, since they boarded in James's house and shared the same cemetery plot later in the century.[31]

John would have spent many days on the porch of his son's house, reading newspapers, corresponding with his extended family perhaps, and reflecting on how far he had come from witnessing slave coffles as a child in rural Virginia, or sitting on Mrs. Taliaferro's stool as a slave boy in Fredericksburg, or even from General Rufus King's mess tent, to retiring in a leafy seaside resort in Massachusetts. He may have avidly followed press reports of ubiquitous blue-gray reunions, monument unveilings, and commemorative writings as the country experienced the simultaneous anniversaries of the Civil War and emancipation. John surely had no doubt where he stood as the vast majority of white Americans embraced a reunion of North and South while forging a system of racial apartheid and nearly erasing the story of black freedom from the nation's mind.[32] He, his wife, his sons, and his growing family of grandchildren had never waited for the emerging Jim Crow society to tell them their place in it.

On February 13, 1918, in the final months of World War I, on the day after his former neighbor Abraham Lincoln's birthday and during the week the black historian Carter G. Woodson had recently declared "Negro History Week," John Washington died in Cohasset of a cerebral hemorrhage and chronic nephritis. He was buried less than a mile up North Main Street in Woodside Cemetery, established in the 1890s as a "low-cost burial alternative" for Cohasset residents. Today an elaborate family gravestone sits directly along the main roadway of the cemetery, which in 2006 was examined by the Massachusetts Historical Commission for possible recognition on the National Register of Historic Places. The cemetery failed all the standards for inclusion on the Register—the quality of funerary art, fame of those interred, and contingency to a historic district. But the commission's report nonetheless noted that the "most notable Cohasset resident buried in Woodside Cemetery is John Washington . . . an escaped slave who served in the Union army." John might have felt flattered by his promotion in rank. But how telling it is that this heretofore unknown former slave would be deemed the most significant among the dead in the working-class cemetery of an upscale Massachusetts town. Americans are finally becoming aware that slavery left indelible marks, large and small, on the national psyche as well as on the American landscape. The Union dead are buried nearly everywhere, and they haunt us still.[33]

———

Wallace Turnage lived a similarly long life, but did not enjoy the same ultimate comforts as John Washington. Just as Turnage had escaped bondage in the face of greater violence and obstacles than Washington, so he continued to live a life defined by struggle even after he was free. Washington's family had middle-class aspirations, but Turnage's family provides a reminder of the travails of the working poor, among which many former slaves were included. Turnage, like Washington, struggled to forge a family in freedom despite tragedy and loss; and he and his brood lived "in unusual evidence," against all the odds.

After the Maryland regiment he served in was disbanded, Turnage resided in Baltimore until 1868. Sometime during this period he married his first wife, Mary, who appears to have died soon thereafter. At the end of the narrative he writes that in Baltimore "I heard from my people and seen them again from which I had been absent from 1860 to 1868." It is unclear how he contacted his mother, Courtney Hart Tyson, and his five half-brothers and sisters in North Carolina, but he must have traveled to the Snow Hill area to visit them. Turnage (like Washington) was among the fortunate freedmen who were able to locate and reunite with their kinfolk in the diaspora caused by slavery and its destruction. In September 1866, Wallace's former owner, Senetty Hart, died; two of her sons had migrated to Texas after the war and one had died in the Confederate army. And in December 1866 Wallace's father, Sylvester Turnage, died in Kinston, North Carolina, just down the road from Snow Hill.[34] In all likelihood

Courtney reported this news to her son. These deaths must have made Turnage resolve to save what he could from his boyhood and retrieve his mother from the Reconstruction South.

By January 1870, Turnage had moved to New York City. In 1871 he opened his first Freedman's Savings Bank account in New York, as well as a second account in Baltimore. Wallace eked out a living waiting on tables at a restaurant called Metropolitan and rented rooms in a tenement at 526 Broome Street, in the heart of what was then known as "Little Africa"—now Soho and Greenwich Village. The area was densely populated and had the largest concentration of African Americans in the city. A major portion of the black population remained in these streets west and south of Washington Square well into the 1880s. Roughly twelve hundred African Americans lived in the neighborhood, including a large number from Virginia known as the "Richmond Negroes." The area was also racially and ethnically diverse, with significant numbers of Italians and Irish occupying tenement buildings with blacks.[35]

By 1890 the center of Manhattan's black population had moved northward into the upper Twenties and lower Thirties on the west side, an area that became, in part, a vice district known as the "Tenderloin," complete with bordellos and nightclubs. And by 1900 black neighborhoods had emerged even farther north in what was called San Juan Hill, West Sixtieth through Sixty-third Streets, now the area of Lincoln Center.[36] Serving as a cook for Captain Turner and other Union soldiers was valuable experience for Turnage, but the twenty-five-year-old surely strained to keep faith in a future in those chaotic, unhealthy streets of lower Manhattan.

To Turnage, a religious country boy and a savvy veteran of the war and five escape attempts, the streets of New York would have been both enticing and psychologically unsettling. Three decades later, in the novel *The Sport of the Gods,* writer Paul Lawrence Dunbar brilliantly captured the experience of a young man's first encounters with New York:

> To the provincial coming to New York for the first time, ignorant and unknown, the city presents a notable mingling of the qualities of cheeriness and gloom. If he have any eye at all for the beautiful, he cannot help experiencing a thrill as he crosses the ferry over the river . . . and catches the first sight of the spires and buildings of New York. If he have the right stuff in him, a something will take possession of him and will grip him . . . every time he returns to the scene and will make him . . . hunger for the place when he is away from it. Later, the lights in the busy streets will bewilder and entice him. He will feel shy and helpless amid the hurrying crowds. A new emotion will take his heart as the people hasten by him,—a feeling of loneliness, almost of grief, that with all these souls about him he knows not one and not one of them cares for him.

Dunbar knew of what he wrote; the young writer nearly died of tuberculosis, a disease that Turnage would come to know all too well. But the seductions of the city could not be resisted:

> After he has passed through the first pangs of strangeness and homesickness, yes, even after he has got beyond the stranger's enthusiasm for the metropolis, the real fever of love for the place will begin to take hold of him. The subtle, insidious wine of New York will begin to intoxicate him.

> Then, if he be wise, he will go away, any place,—yes, he will
> even go over to Jersey.[37]

To Jersey is exactly where Turnage would eventually go, but first
he mixed with the lonely souls in the streets of New York, probably breathing in some of the metropolis's intoxications while
battling against a common laborer's daily emergencies. He also
moved his mother to the city from North Carolina and found a
new wife. He had not survived war and the swamps of Alabama
to let poverty and fate defeat him in New York.

By 1872 Turnage had deposited $125.92 in his bank account
and had taken a job as a janitor, probably in the new seven-story
Equitable Building, New York's first high-rise, which was at the
corner of Cedar and Broadway. At about this time his mother
moved to New York with Wallace's five half-brothers and sisters;
Courtney Hart Tyson lived on Elizabeth Street by 1874 according
to the Freedmen's Bank account of her son, Nelson Hart. In 1874
Turnage worked as a waiter for a William H. Brooks at 218
Thompson Street. He met Sarah Ann Elizabeth Bird, who lived
near his mother; they were married on May 10, 1875, by the Reverend William Spelman at the Abyssinian Baptist Church on
Waverly Place, just west of Washington Square. Approximately
twenty-eight years old, Sarah was probably Brooklyn-born, the
daughter of Francis Bird, a mulatto seaman born in Virginia, and
Frances Bailey. The Birds had lived near the naval yard in the
Williamsburg section of Brooklyn since the 1840s.[38]

By any measure, Turnage and his extended family were urban
poor. Courtney Hart and her five children lived in a sixteen-
family tenement on Cornelia Street, and Wallace and Sarah began
married life in a multifamily building on Thompson Street teem-

ing with children, pathogens, and inadequate ventilation and sanitation. Wallace switched jobs repeatedly from waiter to janitor to "globe man" (a glassblower who made lamps for streetlights), and the family moved frequently as well. Working in service jobs for little more than several dollars per week and often taking in washing, Turnage and his kin struggled to find decent housing and a safe neighborhood, renting wherever they could find the best arrangement. They lived at both 113 Thompson and 13 Minetta, streets made notorious by the writings of the journalist and reformer Jacob Riis in *How the Other Half Lives* and by the young novelist and reporter Stephen Crane. Riis and Crane described plenty of wretchedness and debauchery in these streets, especially on Minetta, a one-block area characterized by "black and tan saloons" (mixed-race drinking establishments) and populated by murderers and muggers with colorful names like "Bloodthirsty," "No-Toe Charley," and "Pop Babcock"—the proprietor of a squalid restaurant. The muckraker Riis made the most of what he called the "moral turpitude of Thompson Street," and Crane found all the violence on Minetta that he desired for his sketches. The district's crime attracted a great deal of police attention, but some respectable black families managed to survive in the area.[39]

The Turnages had moved from this squalid neighborhood before Riis and Crane came to observe it. Suffering from the aftermath of the Panic of 1873, an economic depression that lasted nearly to the end of the decade, and fearing for their health in the lethal disease environment of "Little Africa," the Turnages moved across the Hudson River to Jersey City in 1879. But not before their first two children were born, daughters Ida in 1876 and Sarah in either 1877 or 1878. In Jersey City they rented a house for their

growing family on Van Horne Street in the Lafayette section.
Here, Wallace would live the rest of his life, commuting by ferry
to the various jobs and the community and church life he maintained in Manhattan.[40]

The New York that Turnage vacated strained under the social
weight of severe economic contraction, mass unemployment, political corruption, virulent nativism, racism, rising class tensions,
and violent religious strife between Protestants and Catholics. On
July 12, 1871, an intra-Irish riot resulting from a parade by Protestants through a Catholic neighborhood claimed the lives of sixty-
two New Yorkers, most of them killed by a militia that fired
wildly into the crowds. Wallace could not have belonged to the
racially segregated Workingmen's Association or the trade unions,
even if he had felt inspired by their appeals for broadened liberty
and workers' rights. We do not know how he exercised his Fifteenth Amendment right to vote in these years, but he surely
would not have supported the white supremacist and antidemo-
cratic policies of the New York Democratic Party.[41] Turnage was
what official records always termed a "common laborer"; he strove
daily for basic respect and security, for an hourly wage on which
he and his growing family could survive. He obviously came to
believe that such survival, and residence in a real house, meant
moving to New Jersey.

In *The Black North in 1901*, DuBois mused about well-
meaning whites who sometimes wondered how working-class
blacks could "stand it" under the conditions of poverty and racial
prescription. "The answer is clear and peculiar," he wrote. "They
do not stand it; they withdraw themselves as far as possible from
it into a world of their own. They live and move in a community
of their own kith and kin and shrink quickly and permanently

from those rough edges where contact with the larger life of the city wounds and humiliates them."[42] Turnage knew humiliation all too well, and he both withdrew from and engaged in the life of Manhattan as a black man in a black world partially hidden within a city ruled by warring factions of white men. And at night he retreated across the river to Jersey City.

Turnage appears to have maintained his membership at the Abyssinian Baptist Church, and he joined a fraternal order, the Hamilton Lodge of the Grand United Order of Odd Fellows, chartered in England in 1843. Fraternal orders were crucial community and social organizations in northern black communities. Lodge activities were extremely important to working-class men for whom a proud, supportive, patriarchal association would provide much-needed self-respect and social connection. Turnage and his family might have participated in the Odd Fellows annual parades in lower Manhattan; he could have lifted Sarah's spirits by taking her to lodge-sponsored dances, as well as to the order's picnics and "evening festivals." His own Hamilton Lodge sponsored a masquerade ball in Manhattan in 1887, which drew "Marie Antoinettes," "princesses," "tamborine girls," "Oriental queens," and "Richard IIIs," "Hamlets," and "Romeos" strutting about in costume.[43] Or perhaps Wallace simply confined his lodge activities to church basement meetings and social uplift efforts in the black community.

Apart from his own fraternal order's events, Turnage would surely have been in the "immense multitude" of people looking on during the spectacular parade of Union army veterans down Fifth Avenue on Memorial Day, 1877, proudly displaying pictures of their fallen brethren on wagons and floats as they marched. The former slave would have felt a deep sense of common cause with the seventeen Grand Army of the Republic posts in blue. A magnificent floral display spelled out the word EMANCIPATION across the front of the statue of Abraham Lincoln in Union Square, only a short walk from where Turnage lived; and the black abolitionist and minister Henry Highland Garnet delivered an opening prayer for a ceremony at the Lincoln statue on that beautiful May morning. A year later to the day, Frederick Douglass addressed the racially integrated Abraham Lincoln post of the GAR in Madison Square at Twenty-third Street and Fifth Avenue. How it must have rung true in Turnage's ears to hear the great Douglass denounce the reconciliationist spirit overtaking the country's memory of the Civil War! "There was a right side and a wrong side in the late war," Douglass roared, "that no sentiment ought to cause us to forget." Douglass called his listeners to a higher remembrance, to recall stories and images with the former slaves at the center. The struggle had not been one of mere "sectional character," he asserted. "It was a war of ideas, a battle of principles . . . a war between the old and the new, slavery and freedom, barbarism and civilization." The war was "not a fight," he concluded, "between rapacious birds and ferocious beasts, a mere display of brute courage and endurance, but it was a war between men of thought as well as of action, and in dead earnest for something beyond the battlefield."[44] If Turnage was in that audience, his eyes must have filled with joy and his heart must have pounded with reassurance as he blinked and nodded in agreement. Surely Wallace Turnage knew that his own emancipation and the life he was trying to build were the "something" that Douglass described, the inheritance the great orator and former slave had on that Decoration Day made it possible to see. Even if Turnage was not at the speech, Douglass spoke for him.

During these years Sarah and Wallace needed their spirits raised whenever possible. In June of 1880, not long after moving to Jersey City, Sarah gave birth to twins, Frances (Fanny) and Wallace Jr. The following year William was born, but then a series of family tragedies struck the Turnages. Fanny died of "consumption" (this often meant tuberculosis) at the age of one and was buried in Cypress Hills Cemetery in Brooklyn, in a plot provided through a death-benefit fund by the Hamilton Lodge. The Turnages would bury three more children there over the next seven years. A daughter, Abbie, born in 1883, died of "convulsions" at nine months on September 16, 1884. Only three days later, Wallace Jr., Turnage's namesake, died of "diarrhea and exhaustion" at the tender age of four. On December 10, 1888, Ida, the first-born, died of "consumption" at the age of twelve, leaving only eleven-year-old Sarah, seven-year-old William, and three-year-old Lydia, their seventh and last child, to grow to adulthood. After all these pregnancies, births, losses, and journeys to the Brooklyn cemetery, Sarah died at about this same time (in early 1889) at the age of forty.[45]

A recent study of childhood death rates in late nineteenth-century America named race as the single most decisive statistical factor in high mortality, with "size of place," the "hazards of urban living," and diseases next. Turnage's family was on the wrong side of all four of these measures. In 1900, 18 percent of American children died before the age of five. Mortality rates were considerably higher in large cities, and highest of all for blacks. Approximately 28 percent of black children died before the age of five.[46] In losing four of their seven children, the Turnages had contributed more than their share to the high rates of infant mortality among America's urban working class. Van Horne Street in

Jersey City may have saved Turnage's children from the violence and squalor of Minetta in the West Village, but not from the deadly diseases of urban life.

We cannot be sure when Wallace wrote his narrative. From indications in the text we know that he started and stopped, and composed different pieces at different times. He is not likely to have had much time to write until well after he moved to Jersey City; indeed, the religious tone of the document, especially its ending, may have emerged in the wake of this litany of personal tragedies. As his children continued to die, at least he had a story to leave those who remained. On the final page of his narrative, Turnage writes a "coda," begging the reader to not "take it [the narrative] for a novel, nor a fable, but a reality of facts." Then he proceeds to a prayerful expression of his Christian faith, not a fable indeed but rather a psalm of freedom: "Oh that I may when done with this toilsome world even with three times the difficulties and persecutions that I met with in obtaining my temporal freedom, by God's assistance reach that Blistful abode, and triumph over the enemies of my soul at last. That will be a day of joy to me, which no tongue can express, for I will then be free indeed." In the wake of his many personal losses, a freedom beyond the temporal and above all the graves he had covered over in Cypress Hills Cemetery meant a great deal to him. Wallace wanted his "book" to "show the goodness of God." He wanted to appear grateful; but he also felt chastened. Then he borrows a line directly from the book of Acts, where the apostle Peter appeals to Philip to repent and give his heart to God: "When I prayed to him for my soul's freedom, he for Christ's sake freed my soul from the gall of bitterness and the bond of my iniquity." In the biblical story Philip goes on to encounter an Ethiopian eunuch, whom after great spiritual travail and the reading of scripture he baptizes in a river. The religious ending of Turnage's narrative follows from his remembering that desperate baptism he experienced in the Foul River and in Mobile Bay off the coast of Alabama in 1864. Perhaps he also felt a kinship with Jesus, who was forced to drink "gall" on the cross.[47]

Turnage had one more burial to attend to. On October 6, 1898, his mother died of "acute bronchitis" at the age of sixty-seven. She had lived her last years in a tenement on the 200 block of West Sixtieth Street in Manhattan. She had worked as a nurse-maid, a cook, and a washerwoman during her twenty-six years in New York.[48] The upheavals and changes in Courtney's life were of epic proportions—from slavery to freedom, from a tobacco farm to a New York tenement. She is saved from anonymity; her son and the "unusual evidence" in public records saw to that. But her story, like her son's so ordinary and yet so extraordinary, makes us wish we could know her even better.

Turnage buried four children, two wives, and his mother in the bittersweet free soil of the North. Within months of the death of his second wife, he remarried a third time, to Sarah Bohannah on November 13, 1889, again at Abyssinian Baptist near Washington Square in Greenwich Village. This second Sarah was born in Louisa County, Virginia, probably in 1845. She was the daughter of Mitchell Bohannah, presumably a former slave who had migrated to Washington, D.C., where he drove a milk wagon after the war. Sarah had lived in New York since the late 1860s, just like Turnage, and was working as a servant. Her new husband dearly needed her help with his three small children; on the marriage certificate he listed his occupation as "general jobber," a middle-man who bought goods from retailers and sold them on the streets at cut rates.[49]

Sarah moved into the house on Van Horne Street while Turnage continued to ride the ferry into Manhattan, where he eventually worked as a watchman in various buildings in the Wall Street area. After 1885 he would have seen the Statue of Liberty every day, and he would have witnessed the large ships arriving at Ellis Island filled with European immigrants. He no doubt competed for work with some of those Russians, Slavs, Germans, Jews, Italians, and Hungarians, fellow migrants all. As they mingled, one wonders if the light-skinned African American might have told any of his own story of liberty to those new Americans. Perhaps feeling the absence of his own turbulent, stunning story amidst the great immigrant tumult of New York gave him yet another prod to write his narrative. He faced one of those agonies of the color line in modern America about which Ralph Ellison writes so insightfully: "Everyone wants to tell us what a Negro is, yet few wish, even in a joke, to be one. But if you would tell me who I am, at least take the trouble to discover what I have been."[50]

On October 4, 1916, at the age of seventy, Turnage died of kidney disease and was buried next to his children and his second wife in Cypress Hills Cemetery. His long struggle to provide them a good stake in life—symbolized in part by how dapper he appears in his photographs—was now over. In the wake of Turnage's death, his third wife repeatedly and unsuccessfully applied for a widow's pension based on his service in the Civil War. In a 1919 letter to the Pension Bureau, she mistakenly claims Turnage had enlisted and served in a North Carolina regiment that she could not name for three years during the war. At least twice she received respectful letters in reply from a pension commissioner, informing her that she must supply "the letter of the company and the number of the regiment in which your late husband served and the place of his enlistment and discharge." No such documentation existed, of course. Sarah may have known that, but there was no harm in her trying. At the root of this pension appeal is the sad irony that no "freedom dues" were ever paid to American slaves. Turnage had rendered his service in blood along the war-torn roads of Mississippi, and his discharge might be considered the whipping post of the slave jail in Mobile. Had it been possible or legal, Sarah Turnage might have listed the Foul River estuary, or the waters of Mobile Bay, or the interior parade ground of Fort Gaines as the place of Wallace's enlistment. Like so many thousands of African Americans who broke with their past and entered a new life in the Civil War, Turnage had no pension file record where, under "statement of . . . service," it might simply have read: "Slavery."[51]

Turnage's children lived lives of economic and personal struggle. William had become a shipping clerk and by 1910 had married and moved to Brooklyn, where he later worked as a truck driver. In 1918 he registered for the military draft and listed his occupation differently, this time as a carpenter. Turnage's only surviving son died at age forty-five in 1928 of cirrhosis of the liver, leaving no children to carry on his name. On his death certificate, William was identified as "white."[52] Although he had passed for white and tried to escape the burden of race, he was still bound by class.

As late as 1910 the Turnages' youngest daughter, Lydia, lived at home with her parents in Jersey City and contributed to the household income as a dressmaker. At some point in the 1930s she

married a man fifteen years younger, Thomas Connolly, the son
of Irish immigrants, and followed him to Greenwich, Connecticut, where he worked as a hotel bellman. Like her older brother
before her, Lydia passed for white; she frequently described herself as "Portugee" to explain her tan complexion. She outlived her
husband by twenty years, dying of congestive heart failure in a
nursing home in October of 1984 at the age of ninety-nine.[53]

Through all those years, Lydia carefully kept her father's
emancipation narrative, which he had written on blue-lined paper
in a stationery book purchased from Corlies, Macy & Co., Sta-
tioners, on Nassau Street in lower Manhattan. The narrative was
lovingly preserved in a black clamshell box, which Lydia or someone had labeled ADVENTURES AND PERSECUTIONS 1860–1865.
Since she was passing as white, it is possible she never showed her
father's writings or his photographs to anyone, even her husband.
When Lydia died, her only friend appears to have been a former
neighbor, Gladys Watt, who collected her effects, including the
box with Wallace Turnage's narrative and accompanying photographs. Nineteen years later Mrs. Watt, after watching a documentary about slave narratives, contacted the Historical Society
of the Town of Greenwich and offered to donate her former
neighbor's special heirloom. On the back of a stunning photograph of Turnage, Mrs. Watt wrote an inscription: "Wallace Turnage, born into slavery, son of slaveowner & slave. He married a
white woman and fathered Lydia Connolly . . . who never told
me of her father's background but always praised him as a 'wonderful man.'" In parenthesis, Watt adds: "never mentioned her
mother." As though Watt now knew that she had something the
larger world might want to see, she then wrote: "Among Lydia's

possessions when she died was Wallace Turnage's hand-written story of his 'adventures' during the Civil War."[54]

From these shards of unusual evidence we begin to see a remarkable American story of race, slavery, freedom, tragedy, and survival. We also see glimpses of a story of guarded privacy and proud heritage, of the will to be known that becomes the will to write despite the barriers of status and education. Lydia had saved a daughter's hidden treasure, probably assuming no one would ever care to see it; and if they did, she might have to explain her racial identity. Knowingly or not, she also saved the narrative for the day when a broader world would discover it. Liberty in many forms now emerges from that black box in which Lydia hid her father's story and a piece of her own identity.

CHAPTER FOUR

The Logic and the Trump of Jubilee

What but the love of freedom could bring these . . . people
hither?

—HARRIET JACOBS AT A CONTRABAND CAMP,
ALEXANDRIA, VIRGINIA, 1863

Few events in American history match the drama and significance of emancipation. On Emancipation Day, January 1,
1863, "jubilee meetings" occurred all over black America, North and
South, among nameless ordinary slaves as well as famous abolitionists who had labored for decades to see this moment. At Tremont
Temple in Boston, a huge gathering of blacks and whites met from
morning until night in the beautifully adorned hall, waiting for the
news that Lincoln had signed the fateful Emancipation Proclamation. They were genuinely concerned that something might go
awry; the preliminary proclamation had been issued in September
1862, inspired by a mixture of military necessity, a desire to give the
war a new moral purpose for a war-weary North, and a drive to
thwart the possible intervention on the side of the Confederacy by
a politically divided Great Britain. Britain had ended the slave trade
and enacted emancipation in its empire decades earlier; if the
American war became openly antislavery and not merely a test of

Southerners' right to secede, those British editors, textile mill owners, and aristocrats who favored the South would lose credibility.

Numerous free black leaders spoke at intervals during the day as the growing crowd assembled at Tremont Temple: the attorney John Rock, the minister and former slave John Sella Martin, the orator and women's suffragist Anna Dickinson, and the slave narrative author and historian William Wells Brown. The Bostonian activist and writer William Cooper Nell was the presiding officer. The creator of *Uncle Tom's Cabin*, Harriet Beecher Stowe, looked on prominently from the first row of the balcony. The most famous of all black voices, Frederick Douglass, gave a concluding speech during the afternoon session that was punctuated by many cries of "Amen!" Smiles of anticipation prevailed all around until early evening, when anxiety gripped the hall because no news had arrived from Washington.[1]

Surely Lincoln had not changed his mind. Or had he? Serious concern spread through the mass of people in the great hall as 9 P.M. came and passed. Then a runner arrived from the telegraph office with the news. "It is coming!" he shouted. "It is on the wires!" A speaker tried to read the text of the Emancipation Proclamation as it arrived, but great jubilation engulfed the throng, and the legalistic language of the document dissolved in the din of unrestrained shouting and singing. Douglass led the audience in a chorus of a favorite hymn, "Blow Ye the Trumpet Blow." Next, an old black preacher named Rue led the group in "Sound the loud timbel o'er Egypt's dark sea, Jehovah has triumphed, his people are free!"[2] Historical time had collapsed; the biblical exodus and the present moment might now truly become the same story. It did not matter at that moment what the document said; pent-up dreams and emotions exploded in a celebration that lasted all night.

From Massachusetts to Ohio and Michigan, in cities and on isolated farms, and in the South where freedmen labored for Yankee troops and took refuge in contraband camps, people sensed that they had a chance that night to walk into a new history. Some celebrations were formal, with readings and speeches, while others were spontaneous. At a contraband camp in Washington, D.C., a crowd of six hundred black men, women, and children gathered at the superintendent's headquarters to sing and testify through most of that New Year's Eve. In chorus after chorus of "Go Down Moses" they announced the magnitude of their exodus. One newly supplied verse concluded with "Go down, Abraham, away down in Dixie's land, tell Jeff Davis to let my people go!" In between songs and chants, members of the group stood up and told personal stories of their experiences as slaves. At two minutes before midnight, the entire assemblage knelt on the ground in silent prayer. Then they began to celebrate and sing again, this time spontaneously making up a song, "I's a free man," with a resounding chorus of "Forever free! Forever free!" Drunk on joy, another detachment of ex-slave revelers sang the more formal "Song of the Negro Boatman," which includes:

> We pray de Lord; he gib us signs
> Dat some day we be free;
> De norf-wind tell it to de pines,
> De wild duck to de sea;
> We tink it when de church-bell ring,
> We dream it in de dream;
> De rice-bird mean it when he sing,
> De eagle when he scream.[3]

Having lived for four months in the District of Columbia, John and Annie Washington might have participated in that assemblage, marking emancipation in biblical time and with nature's sounds and rhythms. They were already living a kind of dislocated, anxious freedom that the Proclamation now gave legality, gravity, and energy.

In Corinth, Mississippi, at the large contraband camp that had grown up in that Union-occupied city, more than two thousand former slaves gathered for speeches and a reading of the Proclamation. According to a fugitive slave captured a week later by Confederates in the vicinity, some were given pistols and asked to go into the countryside and "recruit" black men to join the Union forces. During these very weeks, Wallace Turnage was in Mobile, awaiting his sale to Collier Minge that would introduce him to city life as a slave. Just two months before he had been captured a fourth time in northern Mississippi as he was desperately trying to reach the earliest contraband camp at Corinth.[4] Turnage's liberation was still very much in doubt at that juncture, and he did not sing spirituals in celebration of Emancipation Day. Throughout the war, emancipation was always a faltering, dangerous, and tragic process as well as a series of moments when the earth shifted.

Many years after the Tremont Temple celebration in Boston, Douglass may have best captured the meaning of Emancipation Day, whenever and however it came, when he said, "It was not logic, but the trump of jubilee, which everybody wanted to hear. We were waiting and listening as for a bolt from the sky, which should rend the fetters of four millions of slaves; we were watching as it were, by the dim light of stars, for the dawn of a new day;

we were longing for the answer to the agonizing prayers of centuries. Remembering those in bonds as bound with them, we wanted to join in the shout for freedom, and in the anthem of the redeemed."[5] As a former fugitive slave himself who had endured the perils of escape, Douglass's words were a prayerful remembering of all the John Washingtons living in shanties with young wives and newborn babies, or the Wallace Turnages in their Alabama chains. Both fugitives were charting the logic of the new story of black freedom; with any luck, the anthems might come later.

———

American emancipation was always a complex interplay between at least four factors: the geographical course of the war; the size of the slave population in any given region; the policies enforced at any given time by the Union and Confederate governments through their military forces; and the volition of slaves themselves in seizing their moments to embrace a reasonable chance for freedom. Turnage's and Washington's narratives throw into bold relief and confirm the significance of each of these factors. To the perennial question—who freed the slaves, Lincoln or blacks themselves?—the Turnage and Washington stories answer conclusively that it was *both*.[6] Without the Union armies and navies, neither man would have achieved freedom when he did. But they never would have gained their freedom without their own courageous initiative, either.

The emancipation policies of the federal government evolved with much less certitude than the music and poetry of jubilee days might imply. During the first year of the war, the Union military forces operated with an official policy of exclusion ("denial of asylum"), turning away fugitives who reached the front lines. The official reason for the war was to restore the Union, not to uproot slavery and cause what Lincoln later described as a "remorseless revolutionary struggle." But events steadily overtook the initial policy as fugitive slaves began to flee in large numbers to Union lines in Virginia, in Tennessee, and along the Southern coast. In March of 1861, even before the firing on Fort Sumter, eight runaways arrived at Fort Pickens, a Union garrison on the Florida coast, "entertaining the idea," wrote the Federal commander, that United States troops "were placed here to protect them and grant them their freedom."[7] Appearing only as isolated ripples at first, such audacious blacks who made claims on Union authority would in time become an unstoppable wave of humanity.

In May 1861 at Fortress Monroe in Virginia, the ambitious politician-general Benjamin F. Butler, a Democrat and no abolitionist, declared the slaves who entered his lines "contraband of war." The idea of slaves as confiscated enemy property caught on among the troops and in the popular imagination. Since some of the very slaves entering his lines had helped build Confederate batteries nearby, Butler concluded that it was "a measure of necessity to deprive their masters of their services."[8] When word got out, men, women, and children began to stream into Butler's camp. An evolving emancipation policy took hold with this principle of "contraband" slaves, and a word with new meaning born of the war entered the American lexicon.

The federal legislature soon followed Butler's lead. In early August 1861, striking a balance between legality and military necessity, Congress passed and Lincoln signed the First Confiscation Act, allowing for the seizure of all Confederate property used to aid the war effort. The enactment's definition of property specifically included slaves "employed in or upon any fort, navy yard, dock, armory, ship, entrenchment, or in any military or naval service."9 Although not yet technically freed by this law, the slaves of rebel masters came under its purview.

All Union officers, however, were not friendly liberators of slaves. In the winter of 1861–62, the official stance of the Union armies and the Lincoln administration was conflicted: The policy required exclusion of fugitives when the slaveholders were deemed "loyal," and employment as contrabands only when the masters were judged "disloyal." Such tenuous determinations were left to military commanders on the ground, most of whom hoped to be neither slave catcher nor slave stealer if they could help it. Some slaveholders in coastal Florida declared their loyalty to the Union and were paid by the Federal troops who employed their slaves. In August 1861 General John C. Fremont proclaimed martial law in Missouri and declared free the slaves of all rebel masters in the loyal, unseceded state. Worried about the effect of such a bold move on Kentucky and other border slave states still in the Union, Lincoln promptly countermanded Fremont's proclamation and ordered the general to adhere to the First Confiscation Act. Such increasingly ambiguous and unworkable policies caused considerable dissension in the Union ranks, especially between abolitionist and proslavery officers. Northern regiments in the Maryland and northern Virginia theater faced near mutinies over the issue of returning fugitive slaves, some of whom had become servants to Yankee officers. A lieutenant in the Tenth Massachusetts regiment, Charles Brewster, ushered his personal servant, a fugitive named David, out into the woods with a bundle of clothing and best wishes rather than return him to his waiting master at camp headquarters.10

Soon federal legal and military policy would enable rather than restrict the liberation of slaves. By the spring and summer of 1862, Congress took the lead on emancipation policy. In April it abolished slavery in the District of Columbia, provided compensation to the owners of the 3,100 bondsmen still in the District, and allocated a large sum of money for the possible colonization of freed blacks abroad. The Union troops who received John Washington into their ranks on the north side of the Rappahannock River and told him he was "free" had invoked this freshly enacted provision. The law passed decisively in Congress despite bitter debate over the fundamental question of whether slaves were "property" or human beings with "rights," and therefore protected from any Congressional authority over confiscation. It marked the first time the federal government had enacted immediate emancipaton of slaves. The bill also achieved passage in the face of resistance and anxiety on the part of white Washingtonians who feared what the conservative newspaper *The Evening Star* called "a social and political revolution in our midst." But the genie of emancipation was now out of the bottle with this telling precedent for measures yet to follow.[11]

As for the foreign removal of emancipated blacks, the Lincoln administration pursued a variety of schemes for Central American and Caribbean colonization during the first three years of the war. Indeed, members of Lincoln's cabinet tried to persuade Frederick Douglass to be their colonization czar and lead the effort to recruit blacks into the very plan he so vehemently denounced. In answering Postmaster General Montgomery Blair's entreaty, the angry Douglass condemned colonization plans as proslavery theory in disguise. "Slavery has a lease on life given it by colonization," he maintained. "The whole scheme becomes an opiate to the troubled conscience of the nation and barricades . . . the nat-ural course of freedom to the slave."[12] Douglass saw to the heart of slavery's centrality in this war. A small, short war would never have ended slavery in the 1860s in America; but a big, long one would have to in order for the nation to survive. By late 1862 Douglass felt encouraged about emancipation. However contra-dictory the motives or conflicting the changing policies, he re-marked, "It is really wonderful how all efforts to evade, postpone, and prevent its coming, have been mocked and defied by the stu-pendous sweep of events."[13] Those without power almost always have to put their faith in some logic of "events" by both passive expectation and direct action.

In June 1862 Congress abolished slavery in the Western terri-tories, sweeping much political history and the war's overriding cause into one marvelously ironic heap of refuse. The Dred Scott decision now gasped for life, the infamous restriction of citizen-ship to white men alone its only surviving feature. As Union armies advanced, their demands for black laborers increased markedly. The huge numbers of freedmen entering Union lines, especially as the war swept into the dense slave regions of the Mississippi Valley, began to make a mockery of the exclusion pol-icy. In this ever-widening conflagration, black people were now a human asset that both sides needed, and they forced a new moral economy on each side's leadership. Lincoln himself met with loyal border state representatives from Kentucky and Delaware, urging them to adopt gradual, compensated emancipation plans. Their rejection of the president's overtures—even from Delaware, which had fewer than eighteen hundred slaves—forced Lincoln toward a more direct assault on slavery itself. At a meeting in July, he told recalcitrant border-state congressmen that slavery's days were numbered; it would now dissolve, he warned, "by mere friction and abrasion—by the mere incidents of war."[14] With such a soft word as "abrasion" Lincoln eased up to the reality of events—the war would destroy slavery, and he would soon take a leading role in its abolition. Meanwhile, during all these policy battles in Washington and out on the various fronts, Wallace Turnage embarked on his fourth attempt to escape into Mississippi and toward whatever Yankee army he could find. Uninformed but not unaffected by policy debates, Turnage did not wait for events.

In July 1862 Congress passed the Second Confiscation Act, which explicitly freed slaves of all persons "in rebellion," declared blacks "forever free of their servitude," and excluded no part of the slaveholding South. In effect, this new confiscation measure freed all fugitives who came under Union military authority in any occupied enemy territory. Although this law encouraged colonization to some "tropical" land, it also suggested the enlistment of former slaves as soldiers. The Militia Act, passed the same day, July 17, explicitly provided for employment of blacks in "any military or naval service for which they may be found competent."[15] It was under this legal framework that John Washington worked as a mess servant for General Rufus King's staff during the campaign that ended in the disaster at Second Manassas. He, too, had not waited for events before reaping the benefits of new policies and willing Union officers.

These legislative measures provided a public and legal backdrop for Lincoln's subsequent Emancipation Proclamation, issued in two parts as executive orders that were maneuvered through a recalcitrant cabinet and politically calculated to shape Northern morale, prevent foreign intervention (especially British), and keep the remaining four slaveholding border states in the Union.

Increasingly disappointed at the refusal of Southern unionists and border-state leaders to acquiesce to compensated, gradual emancipation, Lincoln remarked privately to two of his cabinet members on June 13, 1862: "We must free the slaves or be ourselves subdued." But Lincoln had long considered slavery to be an evil that ultimately had to be eliminated in America. Now that he was a president with enormous power conducting an increasingly total war for national survival, the questions of emancipation were how, when, and with what imagined consequences. Lincoln was a gradualist by temperament about social change; he had long objected to the rhetoric and tactics of radical abolitionists. He was also an enormously skilled politician, sensitive to timing and the tenor of the public mind, if he could read it in such turbulent times. But this man of the prairie, who grew up fond of racist Negro dialect stories and struggled to imagine an American future of racial equality, nevertheless had a moral compass guiding him in this ultimate crisis of war and social revolution. In the summer of 1862, Lincoln had decided to free the slaves in order to win the war and, in so doing, broaden the very idea of American liberty.[16] But he desperately needed a military success onto which he could graft this plan to destroy the South's labor system and social fabric.

In these very tense weeks, while awaiting the right moment to launch emancipation, Lincoln invited the delegation of African Americans to the White House to hear his views on colonization. That encounter with black leaders was decidedly his worst hour in race relations. Historians have long debated Lincoln's real intentions on colonization. Was he merely trying to condition public opinion for the emancipation that was to follow? Was he pandering to racism and trying to satisfy the divided constituencies of his own party on this sensitive issue? Was he hoping some blacks would begin a process of removal in order to provide a long-term model for how a racially divided postwar society might re-imagine itself? Or did he sincerely hope, along with members of his administration, that blacks in large numbers would voluntarily leave the United States? Lincoln often seemed the embodiment of contradiction and paradox. W. E. B. DuBois was drawn to this quality in Lincoln's character, and he saw it as pivotal on the question of emancipation. "I love him [Lincoln]," DuBois wrote in 1922, "not because he was perfect, but because he was not and yet triumphed." Lincoln could insult blacks and honor them, offer them no American future one month and the next month issue a document that helped give them a new history. "There was something left," DuBois said of Lincoln, "so that at the crisis he was big enough to be inconsistent—cruel, merciful, peace-loving, a fighter . . . protecting slavery, and freeing slaves. He was a man—a big, inconsistent, brave man."[17]

This ambiguous Lincoln kept the public guessing about his deepest priorities, all the while planning privately to launch a bombshell that would reorient the purpose and scale of the war. A week after the meeting with the black delegation, he wrote his famous reply to the editor of the *New York Tribune*, Horace Greeley, who had just chastised Lincoln for his inaction on emancipation. Greeley's letter, called "The Prayer of Twenty Millions," claimed to speak for vast numbers of Northerners who were "sorely disappointed by the policy you seem to be pursuing with regard to the slaves of Rebels." Surprisingly, Lincoln took Greeley's bait, and their correspondence became a public sensation. "My paramount object in this struggle is to save the Union," wrote the president, "and it is not either to save or to destroy slavery. If I could save the Union without freeing any slave I would do it, and if I could save it by freeing all the slaves I would do it; and if I could save it by freeing some and leaving others alone I would also do that. What I do about slavery, and the colored race, I do because I believe it helps to save the Union; and what I forbear, I forbear because I do not believe it would help save the Union." Lincoln had said just enough to invoke varying interpretations of his intentions. But given the course of events in the coming months, this statement was honest on both sides of the semicolons. As he ended his public letter to Greeley, Lincoln made the telling distinction between what he perceived as his "official duty" and his "personal wish that all men everywhere could be free." And his subsequent explanation of the Greeley letter confirms that timing, the situation of the war, and Northern morale were paramount in Lincoln's mind. In responding to Greeley, Lincoln said he wished to be clear that "he would proclaim freedom to the slave just as soon as he felt assured he could do it effectively; that the people would stand by him, and that, by doing so, he could strengthen the Union cause." Never before in American history had pragmatism and idealism combined with such potency in the interest of the liberation of black people.[18]

In the immediate aftermath of the battle of Antietam—the bloodiest day of the war, fought in the cornfields, meadows, and creek bottoms of western Maryland on September 17, 1862—Lincoln's official and personal motivations finally mingled as one. The Confederate army, commanded by Robert E. Lee, though not decisively defeated, was halted in its invasion of the North and forced to retreat southward below the Potomac River. At the ghastly cost of more than five thousand dead and over eighteen thousand wounded in eight hours of fighting, Lincoln and the federal government had achieved enough of a military success for the president to announce the Preliminary Emancipation Proclamation five days later. As he watched the new waves of wounded soldiers and Maryland freedmen stream into the capital, John Washington observed these stirring events while struggling to house his young family. Although he felt free, he surely knew that his ultimate liberation and that of his wife and newborn son depended on a Union military victory.

In the Preliminary Proclamation, Lincoln announced his intentions to free all slaves in those states still "in rebellion" on January 1, 1863. He promised compensation for the loss of slave property and aid in organizing black colonization for any state that would enact "immediate or gradual" abolition. This was, presumably, a carrot to the border slave states and to any Confederate state that might in the ensuing three months give up both the war and slavery in one improbable act. And, perhaps most important, Lincoln pledged the full "military and naval authority" of the United States to "recognize and maintain" all blacks in "any efforts they may make for their actual freedom."[19] Soldiers and sailors now had not only the duty to fight but also to free slaves. Critics of the Proclamation could rightly claim that this was a *military* emancipation with restricted legal scope, but those words "actual freedom" were what rung in the ears of African Americans.

In his Annual Message to Congress, one month before the Proclamation was to take effect, Lincoln laid out once again an elaborate defense of compensation, colonization, and gradual abolition, suggesting the process would take thirty-seven years. He argued that compensated emancipation, if it could shorten the war, would save the nation money. And he unequivocally gave his support for the deportation of blacks if it was voluntary. But one thing was now amply clear: Slavery was to die if the nation were to live. "Without slavery," Lincoln wrote to Congress, "the rebellion could never have existed; without slavery it could not continue." In his famous conclusion to the Annual Message, Lincoln signaled the broadening of the war's meaning to all factions, especially those whites he knew would despise the Proclamation: "We—even we here—hold the power, and bear the responsibility. In giving freedom to the slave, we assure freedom to the free—honorable alike in what we give and what we preserve. We shall nobly save, or meanly lose, the last best hope of earth."[20] Americans have never stopped debating the extended meanings of those few sentences. In deeds inspired by such eloquent words, freedom would soon be given; but John Washington had already seized his liberty in Virginia and Wallace Turnage was desperately pursuing his against all odds in the harrowing physical and psychological terrain of Alabama and Mississippi.

On New Year's Day, 1863, Lincoln signed the great document, which he would later describe as "the central act of my administration, and the greatest event of the nineteenth century." Throughout the war zones, Union forces received thousands of copies of the Proclamation to distribute as they pushed farther into the Confederate heartland. Despite the apparent limitations of the Proclamation—critics then and since have argued that it applied only to those regions over which Federal forces exercised no immediate power—the resounding phrases "then, thenceforward, and forever free" and "henceforward shall be free" entered the lexicon of American civil religion. Lincoln wrote the Proclamation in the language of a legal brief partly because he anticipated constitutional and court challenges from former slaveholders demanding the restitution of their property once the war was over. He was

quite open in his justification of the enactment "upon military necessity," which for some has always tarnished the document's
moral standing.[21]

But the Proclamation's impact on the war and on slaves in the
South was no less real for all its legal dullness and Constitutional
caution. Lincoln had issued a formidable executive order for all
the world to know that now every forward movement of the
Union armies and navies would be a liberating step. Since it came
from the commander-in-chief, the Proclamation carried more
weight among Union soldiers than did the Second Confiscation
Act, and it reduced the ambiguity in federal policy toward escaped slaves. It strengthened pro-Union support in Europe, especially in Great Britain. More than anything, the Proclamation,
despite its careful language, provided an open invitation to slaves
to flee at every opportunity to Union forces. And Lincoln's decree
finally officially sanctioned the enlistment of black soldiers and
sailors. By the summer of 1864 on Dauphin Island, when General
Granger offered Wallace Turnage the choice to enlist or perform
as a camp servant, such actions were almost routine; blacks had
served heroically in Union blue for more than a year despite brutal discrimination.[22]

<hr/>

Across the South, the slave grapevine spread the news of emancipation. A group of Confederate prisoners of war were asked in
1863 about the effects of the Emancipation Proclamation. Many
admitted that "their negroes gave them their first information of
the proclamation." A South Carolina freedman may have left the
simplest explanation for how slaves gained knowledge of the
Proclamation as fast as their masters did. "We'se can't read," he

said, "but we'se can listen." And in Kentucky in the fall of 1862, an Ohio Union soldier wrote to his hometown newspaper that the slaves his regiment encountered were "very anxious to know every step" taken by Yankee troops against slavery. "I tell you," he said, "they are not so ignorant of political matters as some suppose."[23] Turnage and Washington, of course, were both literate and did not have to discern the state of affairs by listening alone. Washington appears to have had full access to Fredericksburg newspapers and was well informed about the course of the war all around him. Turnage, however, in the rural cotton belt of Alabama and Mississippi, did not likely learn from newspapers as much as from word-of-mouth among his fellow slaves and even from his master and overseers. He may indeed have read newspapers in Mobile by 1863, and he certainly knew where the Union armies were in northern Mississippi—those troops that drew him to one desperate flight after another like a gravitational pull he could not resist.

As the war grew in scale in 1862, Southern slaveholders worried about the loyalty and stability of their slaves, as well as about the essential security of their communities and households. The very thing they were fighting to preserve—a racial order based on a slave society and a plantation economy—now threatened to unravel. Charles Colcock Jones, a prominent coastal Georgia planter, wrote grave letters to his family urging them to exercise extreme measures to prevent "absconding Negroes" from "escaping to the enemy at the coast." He warns his son to fear those "traitors who may pilot an enemy into your bedchamber." If the runaways were not stopped, "the Negro property of the county would be of little value." Modern skeptics who still insist that slavery was not the root cause of the war need only read letters such as Jones's. He named the problem and described the slow revolution of emancipation rising in his midst. "Some Negroes (not many) have run away and gone to the enemy," Jones informs his aunt. "How extensive the matter may become remains to be seen. The temptation of change, the promise of freedom, and of pay for labor, is more than most can stand . . . The safest plan is to put them beyond the reach of the temptation . . . by leaving no boats in the water and by keeping guards along the rivers." Jones unwittingly reveals that he understands exactly why slaves would take to boats to fulfill the "temptation" of freedom. No boats! If only he could keep boats away from his slaves, Jones dreams, he could sustain the "hope" that his "people . . . would continue faithful."[24]

As slaves began to flee their masters, the plantation legend of the "faithful slave" exploded in the faces of slaveholders. Many were astonished and expressed disbelief, while others admitted the reality they observed. "As to the idea of the faithful servant," wrote Catherine Edmonston of North Carolina, "it is all a fiction. I have seen the favorite and most petted negroes the first to leave in every instance." Ella Clanton Thomas of Georgia admitted, "Those we loved best, and who loved us best—as we thought, were the first to leave us." In July 1863 an Alabama slaveholder openly confessed that "the 'faithful slave' is about played out." As the war dragged on, some owners found themselves forced to bargain over their slaves' conditions and even pay wages in some instances in order to bring in a crop at all. In exasperation, Mary Jones of Georgia described her remaining slaves' condition as "perfect anarchy and rebellion . . . they have placed themselves in perfect antagonism to their owners and to all government control. We dare not predict the end of all this." Many slaveholders considered

slaves "demoralized" by the approach of Yankee armies. This usually meant that disobedience, work slowdowns, or some combination of confusion and excitement were thoroughly disrupting
plantation discipline and labor.[25]

In Southern memory, after the war it became necessary to the
Lost Cause legend to believe that most slaves remained loyal,
stayed home on plantations, and protected their white masters'
families. In *The Unvanquished,* William Faulkner offers an alternative memory, suggesting that white Southerners might always
be haunted by the incessant movement of escaped slaves through
the countryside. Bayard Sartoris, the young son of a Confederate
officer, is traveling with his "granny" and his trusted slave sidekick,
Ringo, in a wagon amidst the chaos and destruction of war-torn
Mississippi. Bayard describes the approach of a group of approximately fifty blacks before dawn: "All of a sudden all three of us
were sitting up in the wagon, listening. They were coming up the
road . . . we could hear the feet hurrying, and a kind of panting
murmur. It was not singing exactly; it was not that loud. It was
just a sound, a breathing, a kind of gasping, murmuring chant and
the feet whispering fast in the deep dust. I could hear women too,
and then all of a sudden I began to smell them." The scene becomes at once something mystifying and familiar. "'Niggers,' I
whispered. 'Sh-h-h-h,' I whispered. We couldn't see them and
they did not see us; maybe they didn't even look, just walking fast
in the dark with that panting, hurrying murmuring, going on."
The sun rises, and Bayard and his small party go up the road. "Before it had been like passing through a country where nobody ever
lived; now it was like passing through one where everybody had
died at the same moment."[26] In the war's stark ravages and the
tramping sounds of escaped slaves, the causes and consequences

of the Civil War seem to flow deeply into Bayard Sartoris's senses. Faulkner knew that it was the *unfaithful* slave on those Mississippi roads that haunted the Southern imagination.

Like Washington and Turnage, many slaves waited and watched for their best chance to escape, however uncertain their fate might be. Their ultimate aims may have been universal, but their circumstances, or as Faulkner might say, their "murmurings," were local. They all needed courage, connections, and luck. Octave Johnson was a slave on a plantation in St. James Parish, Louisiana, who ran away to the woods and swamps early in the war. He and a group of thirty, ten of whom were women, remained at large for a year and a half. Had Turnage had the advantages of being with such a group of fugitives, he might have succeeded earlier in Mississippi. "We were four miles in the rear of the plantation house," reported Johnson when interviewed by the American Freedmen's Inquiry Commission in 1864. His band stole food and borrowed matches and other goods from slaves still on the plantation. "We slept on logs and burned cypress leaves to make a smoke and keep away mosquitoes." When hunted by bloodhounds, Johnson's band took to the deeper swamp. They "killed eight of the bloodhounds," the proud freedman claimed, "then we jumped into Bayou Faupron; the dogs followed us and the alligators caught six of them; the alligators preferred dog flesh to personal flesh; we escaped and came to Camp Parapet, where I was first employed in the Commissary's office, then as a servant to Col. Hanks; then I joined his regiment." From "working on task" through surviving in the Louisiana bayous, Octave Johnson found his freedom as a corporal in Company C, Fifteenth Regiment, Corps d'Afrique.[27] From slave to swamp refugee to camp servant to soldier—Johnson got to tell his tale. Wallace had a story even more dramatic than Octave's; no wonder he so dearly wanted to write it down for anyone who would read it.

For the bulk of slaves, the transition from bondage to freedom was not so clear and complete as it was for Octave Johnson. Emancipation was a matter of overt celebration in some places, especially in Southern towns and cities under Union occupation. But what freedom would mean in 1863, how livelihoods would change, how the war would end, how the masters would react (perhaps with wages but maybe with violent retribution), how freedpeople would find protection and food in the war-ravaged and chaotic South, how they would meet potential rent payments, how a peasant population of agricultural laborers attached to the land might now become owners of the land as so many dreamed, and whether they would actually achieve citizenship rights were all urgent and unanswered questions. We do a disservice to the experience of the freedpeople by remembering only their music of spiritual victory and not the physical agony through which they passed.

The actual day that masters announced to their slaves that they were free was remembered by freedpeople with a wide range of feelings and experiences. Some recalled hilarity and dancing, but others remembered it as a sobering, even solemn, time. A former South Carolina slave recollected that on his plantation "some were sorry, some hurt, but a few were silent and glad." James Lucas, a former slave of Confederate president Jefferson Davis in Mississippi, probed the depths of human nature and ambivalence in his description of the day of liberation: "Dey all had diffe'nt ways o' thinkin' 'bout it. Mos'ly though dey was jus' lak me, dey didn' know jus' zackly what it meant. It was jus' somp'n dat de white folks an' slaves all de time talk 'bout. Dat's all. Folks dat ain' never been free don' rightly know de *feel* of bein' free. Dey don' know the meanin' of it." And a former Virginia slave simply recalled "how wild and upset and *dreadful* everything was in them times." Some former slaves no doubt preferred, at least in the immediate wake of slavery, to forget rather than remember. They may have found themselves feeling like the two lead characters in Toni Morrison's landmark novel *Beloved,* when Paul D says to Sethe: "Me and you, we got more yesterday than anybody, we need some kind of tomorrow."[28]

For most slaves, freedom did not come on a particular day; it evolved by a process. For thousands of ex-slaves who followed or searched for Union armies, freedom initially meant life in contraband camps, where they struggled to survive in the face of great hardship, disease, and occasional starvation. In LaGrange, Bolivar, and Memphis in western Tennessee; in Corinth and Holly Springs in northern Mississippi; in "contraband colonies" near New Orleans; in Cairo, Illinois; at Camp Barker in the District of Columbia; on Craney Island near Norfolk, Virginia; at a burgeoning, badly equipped camp in Alexandria, Virginia; and eventually in northern Georgia, coastal North Carolina, and dozens of other places, ex-slaves tried to forge a new life on government rations and by working on labor crews. They received makeshift medical care, often provided by "grannies"—black women who employed home remedies—and Northern female aid workers. The contraband camps also served as the initial entry into the practice of free labor as well as stimulating a new sense of dignity, mobility, identity, and education.[29]

Most white Northerners who witnessed, supervised, or worked in these camps—or who founded schools and freedmen's aid societies and observed weddings and burials among ex-slaves—were stunned by the determination of this exodus despite its hardships. In 1863 each superintendent of a contraband camp in the western theater of the war was asked to respond to a survey about the freedmen occupying his facilities. To the question of the "motives" of the freedmen, the Corinth superintendent tried to capture the range of what he saw: "Can't answer short of 100 pages. Bad treatment—hard times—lack of the comforts of life— prospect of being driven South; the more intelligent because they wish to be free. Generally speak kindly of their masters; none wish to return; many would die first. All delighted with the prospect of freedom, yet all have been kept constantly at some kind of work." The Holly Springs superintendent replied to the motives inquiry very directly: "Universal desire to obtain their freedom." All of the superintendents commented on what seemed to them the remarkable and surprising degree of "intelligence" and desire for "property" among the freedmen. Each superintendent also seemed stunned at the religiosity of the freedmen. At Holly Springs and Memphis, their "leading doctrines of the Bible" were thought "remarkably correct"; and, despite "peculiar notions," they "often pray and sing all night," wrote the Corinth superintendent. As for their "notions of liberty," the Memphis superintendent answered: "Generally correct. They say they have no rights, nor own anything except as their master permits; but being freed, can make their own money and protect their families."[30] Indeed, these reports demonstrate just what a fundamental revolution emancipation had become. How dearly Wallace Turnage must have wished he had made it past the Confederate patrols and through those northern Mississippi roads to Corinth, Holly Springs, or Memphis to find a friendly community and add his young voice to this storehouse of testimony.

Had Turnage reached the Corinth camp safely he would have found one of the largest of all such facilities, although populations were fluid. The camp operated as a model contraband experiment from 1862 until January 1864, achieving some remarkable success in creating stable housing, cooperative farming, increased literacy, burgeoning religious activity, and even a modest profit for the federal government. Word of Corinth's contraband camp spread so widely that a northwest Georgia slaveholder advertised for three of his runaways whom he claimed were en route across Alabama to reach the Mississippi town.[31]

The Corinth camp may have reached a population as high as six thousand at one point, but a March 1862 census reported 3,657 inhabitants: 658 men, 1,440 women, and 1,559 children, virtually all from Mississippi, Alabama, and Tennessee. Black troops were organized to guard the camp, and the local economy hummed on the skills of 36 blacksmiths, 48 carpenters, 180 teamsters, 800 cooks, 80 seamstresses, and 150 laundresses. Four hundred thirty-eight of the men and 1,080 of the women described themselves as married; and initially 120 men and 40 women were literate, with those numbers rapidly rising.[32] Here, in microcosm, was the possibility of black freedom demystified and transformed materially and spiritually. The story of American emancipation is not merely one of transcendent documents or ceremonies or the hypnotic music of God's laborers set free. It surely begins in these camps of refugees, their campfires and conversations pulsating with the feel of humanity and independence.

At Corinth hundreds of freedmen attended prayer meetings almost every night in the summer, and the camp was divided into wards, each with a captain to keep order. By May 1863 three hundred children attended school daily, and one missionary gushed,

"We cannot enter a cabin or tent, but that we see from one to three with books." Like other contraband camps, Corinth became a major recruiting station for black soldiers. At stirring meetings where black men chanted, sang, and pledged their willingness to die for the Union and their own freedom, the one thousand members of the First Alabama Infantry of African Descent were officially mustered into the army as the Fifty-fifth United States Colored Infantry. The regiment gained notoriety for its self-help philosophy carried over from the contraband camp and for the men who carried schoolbooks out on picket duty. Unfortunately, the entire Corinth camp was evacuated in January 1864 on stern orders from General William T. Sherman. Several thousand refugees, many on foot, suffered a midwinter trek of ninety-three miles to Memphis, where they camped first in open fields and then were left largely to their own devices.[33] Wallace Turnage's individual story mirrors the hopes and tragedies of the thousands moving to new beginnings and endings all over the region of the lower Mississippi Valley.

Most freedmen were stalked by poverty wherever they roamed or landed. Wiped out or disbanded by campaigning armies, some makeshift contraband camps did not last long. Some were little more than shantytowns and refugee villages of sod huts. Corinth may have been the exception in having more education and religious meetings than gambling dens, saloons, and brothels. But the freedmen surged on, whatever the odds. And their presence proved simultaneously an aid, obstacle, and bothersome afterthought to the generals making war. In the near panic to move contrabands to Memphis in 1864, a chaplain, John Eaton, observed an episode of chaos: "Their [freedmen's] terror of being left behind made them swarm over the passenger and freight cars, clinging to every available space and even crouching on the roofs."[34]

American popular culture has never depicted these scenes of our Civil War, preferring nostalgia for the common martial glory of the Blue and the Gray to images of starving black children or old men and women with pneumonia languishing in contraband camps.

Back east, Harriet Jacobs, the former fugitive slave and author of the now celebrated narrative *Incidents in the Life of a Slave Girl*, worked heroically among the contraband camps of the U.S. capital. Since Jacobs had spent long periods of time in Washington from June 1862 through the end of the war, she and John Washington surely must have crossed paths, either at church, or working among the burgeoning refugee population, or at freedmen's relief meetings. Even if they did not personally know each other, Jacobs's remarkable writings about the refugee conditions and emancipation process that she observed are a window into the immediate circumstances of John, Annie, and the family they tried to gather around them. In September 1862, the month of John's arrival in the city, Jacobs published a long and revealing article, "Life Among the Contrabands," in William Lloyd Garrison's *Liberator*. At Duff Green's Row, the first of the capital's contraband camps, Jacobs observed "men, women, and children all huddled together . . . in the most pitiable condition . . . sick with measles, diphtheria, scarlet and typhoid fever." Scrambling about in "filthy rags, the little children pine like prison birds for their native element." Ten deaths per day were common at Duff Green's Row. Given these conditions, it is no wonder that John Washington used his literacy and other skills to find paid work and his own housing as fast as possible. He probably found his first jobs at the makeshift employment agency the federal authorities established at this early contraband camp.[35]

By the winter of 1862–63, Jacobs had moved across the Potomac to Alexandria, Virginia, now a huge military depot as well as a collection point for refugees from the bloody campaigns fought within seventy-five miles to the south and west. In her labor as a relief worker, Jacobs seems to have found a new spirit and a new voice, not unlike what Walt Whitman gained from similar work. Despite all the misery, death, and skullduggery she witnessed, the courage of the refugees helped ease some of the "memory of the past" in her own life. In the eyes of those black refugees, whether "desolate" or full of "beaming happiness, the last chain is broken, the accursed blot wiped out."[36] Did she somehow exorcise parts of the memory of a garret prison cell and its accompanying humiliation by rising every day at dawn to try to clothe and shelter freed slaves? Did Jacobs free herself again by defeating some of the racism and bureaucracy found in those refugee camps? By working among the displaced and suffering contrabands, Jacobs seems to have felt like a soldier in the army of the free.

The Alexandria camp, like so many others, was not a pretty sight. Some visitors observed only what their prejudices revealed—a half-clothed mass of uncivilized humanity displaced from their normal condition. Nathaniel Hawthorne observed the camp and thought the freedmen "picturesquely natural" in their "primeval simplicity." They seemed to the conservative New Englander a "kind of creature by themselves, not altogether human . . . akin to the fauns and rustic deities of olden times." Hawthorne's English traveling companion was even more direct in his description of the refugees: "Miserably clothed, footsore, and weary, they crouched in the hot sunlight more like animals than men." Jacobs admitted that many of the freedmen were "degraded by slavery" and "stupid from oppression." But she demanded patience. Her letters are full of refutations of the racism with which most Americans met the prospect of freed slaves. In their poverty she saw no beauty, in their degradation no ancient literary image. In their near nakedness she saw only God's command to human benevolence, in their lack of education a laboratory for enlightenment and virtue. "Anyone who can find an apology for slavery should visit this place, and learn its curse. Here you see them from infancy up to a hundred years old. What but the love of freedom could bring these old people thither?"[37]

She might have been describing John Washington's grandmother who arrived in Washington with her grandson. And given Jacobs's personal experience of resisting her sexually abusive owner and mothering of Harriet Jacobs's two children by another white man, she could hardly resist commenting on all the light-skinned ex-slaves she met in the camps: "Here I looked upon slavery, and felt the curse of their heritage was what is considered the best blood of Virginia." In more ways than one, Virginia's blood was visible everywhere around Washington, D.C., in 1863. Military histories and reference works about the Civil War are replete with the ghastly casualty statistics from battles and campaigns, and it is good that those numbers are learned and rehearsed. But seldom do we learn that in the Washington, D.C., contraband camps approximately eight hundred black refugees died from disease and exposure from October 1862 to March 1863.[38] The costs of freedom were high.

The "logic" of emancipation, to borrow from Douglass, took many forms, often in the long lines of freedmen tramping the roads of Mississippi, Virginia, and all over the war zones of the South. It emerged in legal language and documents crucial to securing the fact as well as the feel of freedom. Eventually it was enshrined in the Thirteenth Amendment. But that logic was also shown in the daily choices made by slaves, by Union officers, by Confederate soldiers and civilians, by the black soldier on picket duty who studied his spelling book, or the freedwoman who desperately wrote letters to find her kinfolk. For the freedmen, the logic sometimes was emotional or naïve, as in a devil-may-care bolt up the railway line or a river bank by a teenager in Mississippi; sometimes it was carefully pragmatic, as in the steps of a young skilled urban man helping to manage a hotel in a Virginia city under military siege.

Any logic in history is made and unmade by events and human will. The Georgia planter John Jones watched the disintegration of slavery with nearly "hopeless despair" in the summer of 1865. He agonized to his sister that managing blacks now was impossible in the immediate wake of the war. He cautioned her that they (as former slaveholders) were "clinging too much to a race who are more than willing to let us go," and a "property . . . which has passed its best days for ease and profit." With self-revelatory clarity, Jones saw the logic and nature of the "great changes" all around him: "The dark, dissolving, disquieting wave of emancipation has broken over this sequestered region. I have been marking its approach for months and watching its influence on our people. It has been like the iceberg, withering and deadening the best sensibilities of master and servant, and fast sundering the domestic ties of years."[39] Jones's candid description provides both an eloquent epitaph for slavery and a reluctant tribute to the epic power of emancipation.

Emancipation in America was a revolution from the bottom up that required power and authority from the top down to give it public gravity and make it secure. Freedom, as Lincoln said, was something given and preserved, but it also, as he himself well understood, had to be taken and endured. And it ultimately was fostered by war and engineered by armies.

In Turnage's corner of the deep South, only small numbers of slaves actually became free by individual or collective wartime flight. This makes Wallace Turnage's escape all the more remarkable. In coastal Georgia and South Carolina, and in many Cotton Belt counties of Alabama or Mississippi, far more slaves managed to escape thanks to a practice of evacuation commonly called "refugeeing." Of Georgia's roughly 34,000 seaboard slaves, for example, a sizeable majority were evacuated as their masters moved them inland away from invading Union forces.[40]

But the refugee strategy was a desperate move on the part of slaveholders. It shattered the stability of agricultural production and plantation life, and as the security of the plantation slowly died, so died slavery. Many bondsmen dreaded these forced removals that caused yet more family separations; but many of them also found paths to the Union lines during their "refugee" journeys. Uprooting is, after all, uprooting. Just as the psychology of the master-slave relationship could not survive the refugee process intact, so did the freedmen's basic attachments to the land and crops unravel. One thing was clear: A refugee planter was a master no more; his or her power crumbled with every desperate step, and their authority was at the mercy of the destructive Federal troops and the will and sudden mobility of their "people."[41]

Still more slaves found opportunities for movement and emancipation in the Confederacy's extensive use of slave labor to work in army camps, build fortifications, and staff munitions factories and iron furnaces. In Virginia, urban tobacco factories like the one where John Washington worked were by far the largest prewar employers of hired slaves. When the war came, cities and towns like Richmond and Lynchburg, as well as the numerous iron furnaces in the Shenandoah Valley, hired black laborers in the thousands from rural areas.[42] This caused accelerated urbanization and physical movement of blacks on an unprecedented scale.

General Robert E. Lee frequently implored Virginia slaveholders to provide black labor to the armies and for military production, and he often received a cold response from anxious owners unwilling to deliver their valuable property to the war effort. Agents of the huge Tredegar Iron Works in Richmond traveled the state, outbidding the smaller furnaces in the valley. Demand for slave labor on wartime railroads also increased as the war surged on. And, of course, slaves who were hired out soon began to flee as the armies approached. One furnace manager described "almost a stampede among the hands" as they fled their posts. Some ironworks operators softened labor conditions and paid "overwork" wages to try to mollify restive slaves. And as more and more blacks were impressed into service with the army, a Virginia planter, John Spiece, issued a telling warning to the Confederate attorney general. He saw "a serious evil in impressing slaves for the service . . . whilst there they get to talking with Union men in disguise, and by that means learn the original cause of the difficulty between the North & South, then return home and inform other negroes."[43] By 1862 that "cause" moved about all over Virginia and in many other regions of the South. In Spiece's stunning honesty, we see both the deepest reason Southerners went to war and the primary "cause" of their undoing.

By the spring of 1865, approximately 474,000 former slaves and free blacks had participated in some form of federally sponsored labor in the Union-occupied South. Approximately 271,000 of that number came under Federal authority in the deep South and 203,000 in the upper South. An unknown number of ex-slaves also had negotiated free-labor arrangements with their former owners. Of the approximately 180,000 black men who served in the Union army and navy—10 percent of the total Federal forces for the war—nearly 80 percent were former slaves recruited in the South. Still other former slaves left the South and became wartime laborers in the North, although their numbers were small due to widespread Northern opposition to the relocation of blacks into the free states. Even antislavery and pro-emancipation news-papers and politicians were wary of the widespread emigration of blacks to northern climes. It was not the freedom of blacks that concerned so many white Northerners, declared a Massachusetts monthly established to promote emancipation, but rather "the proper disposal of the Negro." Some politicians used the fear of a vast freedmen migration to the North to foment support for emancipation of blacks who would stay within the South under military control.[44] Ultimately both Turnage and Washington lived with many of these Yankee prejudices as they made their transition from freedom to an illusive equality.

This statistical breakdown of freedmen within Union author-ity unsurprisingly reflects the most densely populated slave re-gions. The lower Mississippi Valley, including southern Louisiana, accounted for 223,000 of the deep South's total, and it is here where we can count Wallace Turnage's individual escape. By war's end, approximately forty thousand freedmen lived and labored in and around the District of Columbia, and here, of course, John Washington and his kinfolk can be included in some of the numbers.[45] In early 1865, some 600,000 to 700,000 out of the nearly four million African American slaves had reached some form of emancipation.

But they were not merely numbers, as the Washington and Turnage narratives make amply clear. Survivors write their stories, often in the name or spirit of those who do not make it. Perhaps it was a survivor spirit that compelled John and Wallace to pick up their pens and write; they both knew how many others like them were lost to time and the piteously destructive war of liberation. And they certainly understood the impact of dislocation on their own kinfolk.

When thousands of black men entered the Union army or fled to freedom while laboring for the Confederate forces, they often left women and children behind in great hardship, sometimes in sheer destitution, which became worse under new economic arrangements that required rent payments. Some women and children ended up in "regimental villages," contraband camps established for families of black soldiers at the front. In July 1865, Louisiana freedwoman Emily Waters wrote to her husband, who was still on duty with the Union army, begging him to get a furlough and "come home and find a place for us to live in." The joy of freedom was mixed with terrible hardship: "My children are going to school, but I find it very hard to feed them all, and if you cannot come I hope you will send me something to help me get along . . . Come home as soon as you can, and cherish me as ever." The same Louisiana soldier received a subsequent letter from his sister, Alsie Thomas, reporting that "we are in deep trouble—your wife has left Trepagnia and gone to the city and we don't know where or how she is, we have not heard a word from her in four weeks."[46]

The choices and agonies that emancipation wrought are tenderly exhibited in a letter written by John Boston, a Maryland fugitive slave, to his wife, Elizabeth, in January 1862: "[I]t is with grate joy I take this time to let you know Whare I am I am now in Safety in the 14th regiment of Brooklyn this Day I can Adres you thank god as a free man I had a little truble in giting away But as the lord led the Children of Isrel to the land of Canon So he led me to a land Whare Fredom Will rain in spite Of earth and hell . . . I am free from al the Slavers Lash." Such were the exhilarations of freedom and the pains of separation. Boston concluded his letter: "Dear Wife I must Close rest yourself Contented I am free . . . Write my dear Soon . . . Kiss Daniel For me."[47] Surviving sources do not tell us whether Emily Waters ever saw her husband again, or whether the Bostons were ever reunited. But these letters, like Turnage's and Washington's stories, demonstrate the depth with which freedom was embraced and the human pain through which it was achieved.

In 1865 John Washington buried his infant son, Johnnie, in the free soil of the District of Columbia, and Wallace Turnage followed a Maryland regiment back to the now free city of Baltimore. John was trying to make a family and a life as he anguished over how to remember his dead son; Wallace was trying to imagine a new life and a way to locate his lost family in North Carolina. Both were desperate for a sense of home. Both had so much to forget in order to build their new lives. But they would do so in two extraordinary acts of remembering. Perhaps they could never quite realize their tomorrow until they had told the story of their yesterday. The American writer Richard Rodriguez, himself the author of an autobiographical journey across boundaries of nationality and ethnicity, may have best captured the reasons why former fugitive slaves who had found some version of secure freedom might turn to first-person narrative: "Autobiography seems to me appropriate to anyone who has suffered some startling change, a two-life lifetime; to anyone who is able to marvel at the sharp change in his life: I was there once, and now, my God, I am here! (. . . was blind but now I see.)"[48] Turnage and Washington had been lost. But now, as they surely wished, they are found.

A family photograph taken between 1913 and 1918, in Cohasset, Massachusetts. From left: Annie Washington, John Washington, their son James (standing) and his wife Catherine. John and Annie moved to Cohasset and lived in retirement in their son's home until John's death in 1918.

The Alice Jackson Stuart Family Trust

Benjamin Washington (far right), the youngest son of John and Annie, with three football players or coaches. From left: "Ike" Wright, Ed Henderson, "Duck" Gibson. The photo is undated, but it was taken in a major sports stadium in the 1930s or 1940s while Benjamin was commissioner of a black collegiate football conference. *The Alice Jackson Stuart Family Trust*

Evelyn Washington, Benjamin's only child and John's granddaughter, in a prom dress during her senior year of high school, June 1930, Washington, D.C. She became a physical education teacher with a master's degree and later married Leroy Easterly of Zenia, Ohio. Evelyn preserved her grandfather's slave narrative and passed it on to her best friend, Alice Jackson Stuart, who in turn left it to her son, Julian Houston. *The Alice Jackson Stuart Family Trust*

The gravestone of John and Annie Washington at Woodside Cemetery, Cohasset, with son Benjamin standing behind it, August 18, 1937. In 2006 the Massachusetts Historical Commission reported that the "most notable Cohasset resident" buried in Woodside Cemetery was "John Washington ... an escaped slave who served in the Union army." *The Alice Jackson Stuart Family Trust*

From 1858 to 1860 John Washington lived in this small cottage at the back of his owner Mrs. Taliferro's rented house at 409 Hanover Street in Fredericksburg. The cottage still stands today. *Photograph by Alan Zirkle*

Fredericksburg, 1863, with the Rappahannock River in the foreground, where John Washington was baptized. The large building on the riverbank at right is the African Baptist Church where Washington attended services and married Annie Gordon in January 1862. He crossed the Rappahannock to freedom less than a mile upriver from this spot. *Courtesy of the National Park Service*

Princess Anne Street, Fredericksburg, 1863, directly in front of the Farmer's Bank where young Washington lived for a time. This was his view as he ran errands for his mistress or walked to church. *Courtesy of the National Park Service*

In the background of this 1863 photograph of Fredericksburg is the
Alexander & Gibbs tobacco factory (the rectangular building with the
slanting roof) where John Washington worked as a hired slave in 1860.
Courtesy of the Western Reserve Historical Society, Cleveland, Ohio

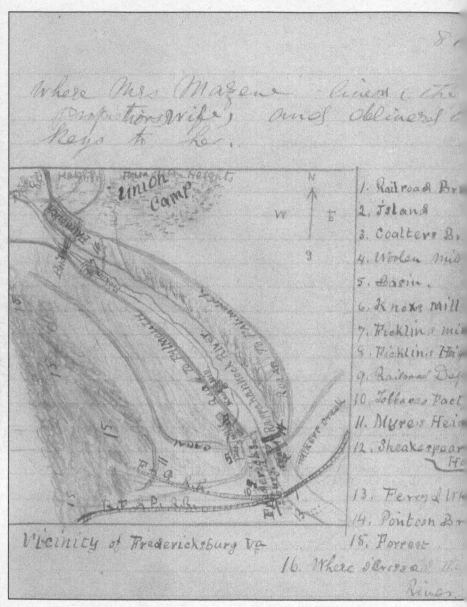

Where Mrs. Magene lived & the
proposition wife, and Olivers &
keys to he.

union Camp

N

W   E

S

1. Railroad Br.
2. Island
3. Coalters Br.
4. Woolen mill
5. Basin.
6. Knoxs Mill
7. Ficklins mi
8. Ficklins Hei
9. Railroad Dep
10. Tobbacco Fact
11. Myres Heig
12. Sheakespear
    Ho
13. Ferry Wr
14. Pontoon Br
15. Forrest
16. Where I crossed the
    River

Vicinity of Fredericksburg Va.

Washington includes this hand-drawn map of Fredericksburg in his
manuscript to help him tell his story of escape, listing all the pertinent sites
in town and along the river. Few if any other slave narratives include such a
personal map of a day of liberation. *The Alice Jackson Stuart Family Trust*

A Mathew Brady photograph of what is today the Mall in Washington, D.C. The unfinished U. S. Capitol building is in the distance, and the original Smithsonian Castle dominates the center. This photograph is circa 1863, shortly after John Washington entered the city. Those may be ambulance wagons with teams of horses along the roadway. Nearly 40,000 refugee freedmen settled in Washington during the war, many of them living initially in contraband camps in Alexandria and elsewhere. *Courtesy of the National Archives*

Former slaves cross the Rappahannock River upstream from Fredericksburg as they travel toward Washington in August 1862. In his narrative John describes witnessing just such caravans of refugee freedmen. *Courtesy of the Library of Congress*

The house at 314 North Main Street, Cohasset, Massachusetts, where John and Annie Washington moved to live with their son James and his family in 1913. John spent many afternoons sitting on the porch and died here at the age of eighty. *Photograph by the author*

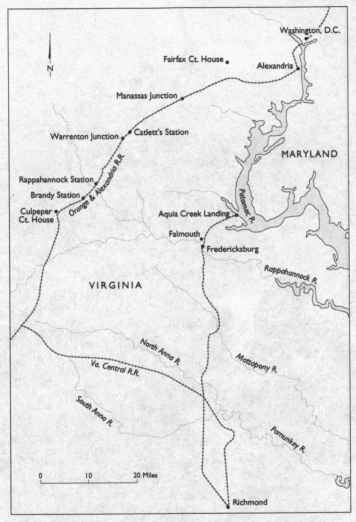

Northern Virginia during wartime. John Washington escaped across the Rappahannock River near Falmouth. On a campaign with the Union army he traveled first west and then north, marching to all the towns along the Orange & Alexandria Railroad, on which he rode into the nation's capital for the first time in late August 1862. *Map by William Nelson*

A lithograph of a runaway slave fleeing through tall swamp grass. Wallace Turnage was chased by dogs and slave catchers in just such settings in Mississippi and Alabama. *Courtesy of the Library of Congress*

The Ballard Hotel on the corner of Franklin and Fourteenth Streets, Richmond, Virginia, in April 1865 after the evacuation of the Confederate capital. Wallace Turnage lived and labored for several months at the Hector Davis slave jail, which stood just behind the Ballard. The Ballard and Exchange hotels and the various auction houses around this intersection were the hub of Richmond's slave-trading business. *Courtesy of the Valentine Richmond History Center*

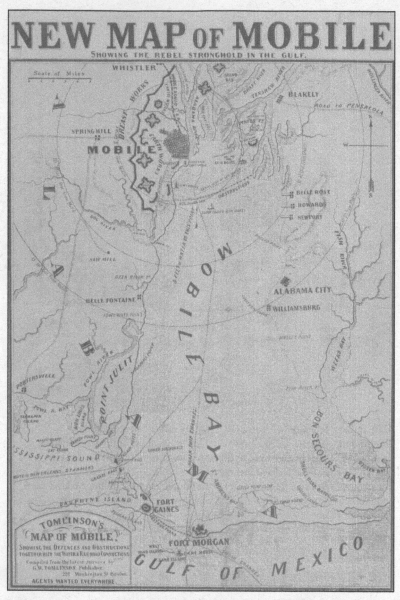

A wartime map of Mobile Bay produced for the U.S. War Department in 1863. The map includes most of the geographic markers Wallace Turnage identifies in his narrative of his escape down the western shore of the bay, a distance of nearly thirty miles from the city of Mobile. *Courtesy of the New York Public Library Map Division*

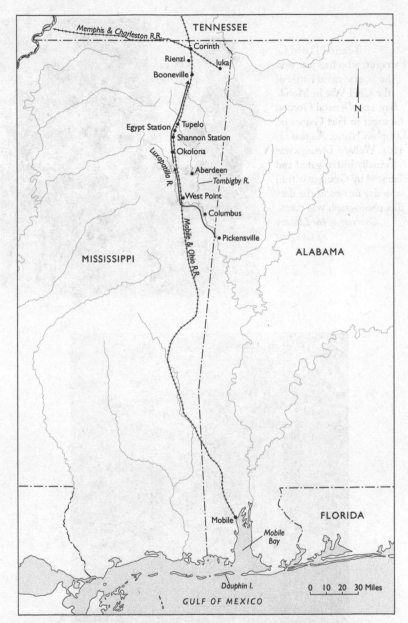

Wartime Alabama and Mississippi. Wallace Turnage ran away four times
from a cotton plantation near Pickensville. Each time he headed west and
north along the route of the Mobile and Ohio Railroad. His goal, he said,
was to "go home," although he really intended to reach the Union forces
that eventually controlled much of northeast Mississippi. The towns
indicated on this map are all mentioned in Wallace's recounting of his
various escape attempts. *Map by William Nelson*

Admiral David G. Farragut, who had just won the largest naval battle of the Civil War in Mobile Bay, and General Gordon Granger in Fort Gaines on Dauphin Island, August 22, 1864. Wallace Turnage was formally interrogated and liberated by Granger within a day or two of the date this photograph was taken. *Courtesy of the Library of Congress*

Minetta Lane, New York City, circa 1922. In the late 1870s Wallace Turnage lived at number 11 (left) on this legendary street, notorious for its poverty, crime, and saloons. *Courtesy of the New York Public Library's Photographs Division*

Lydia Turnage, circa 1905. Wallace Turnage's only surviving
daughter preserved his narrative until her death in 1984. She
lived in Greenwich, Connecticut, and passed for white. *Courtesy
of the Historical Society of the Town of Greenwich, Connecticut*

Wallace Turnage had at least four studio photographs taken with him wearing his finest clothes, a common practice among many working-class people. From top to bottom: circa 1880 in his early thirties; circa late 1880s; and circa mid-1880s. The staged railing and sign might have been inspired by his move to Jersey City. *Courtesy of the Historical Society of the Town of Greenwich, Connecticut*

Author's Note

I HAVE PUBLISHED EACH NARRATIVE IN ITS RAW, ORIGINAL FORM with virtually no changes to the grammar and spelling. Wallace Turnage rarely paused for paragraph breaks, so I have provided them to assist the reader's eye and comprehension. Both men struggled with spelling; occasionally I have provided a word in brackets to indicate the intention of the author. Capitalization is also haphazard in each case and sometimes difficult to discern. When in doubt I have capitalized the first words of sentences. For each narrative I have further provided annotation notes to identify some specific people, places, and concepts. The transcriptions published here were scanned and proofed against the original.

So far as can be determined, neither of these documents was ever edited or refashioned by anyone in any way. We are fortunate that each man had excellent handwriting; and I am personally grateful for the moving experience of seeing both Washington and Turnage struggling on the page to tell their stories. They captured and transported me to 1862 and 1864, and I have come to understand emancipation as never before.

"Memorys of the Past"
John M. Washington

I was born (in Fredericksburg Virginia: May 20th 1838.) a slave to one Thomas P. Ware Sr. who I never had the pleasure of knowing; (I suppose it might have been a doubtful pleasure.) as he died before I was born.

When I was about 2 years of age My Mother (who was also a slave) was hired to one Richard L. Brown in Orange County Virginia. about 37 Miles from Fredericksburg, and I was taken along with her. But I will not promise to Narate the incidents of that Jurney as I did not keep a Diary at that age in a slave state. My reccolections of my early childhood has been no doubt the most plasant of my life. My mother taught me to spell at a very early age (between 4 years and 8)——

When at this time of Life I look back to that time and, all its most vivid reccolections I see myself a small light haired boy (very often passing easily for a white boy.) playing mostly with the white children on the farm, in summer Evening amoung the sweet scented cloverfields after the Butterflys Wading the Brooks and with pin hook and Line startling the spring troute.

Now in the Great Forrest surrounded with dogs in quest of the Hare and Opussum, often on the top of some neighbouring mountain trying with my young eyes, to get a view of Fredericksburg in the distant about 37 miles Eastward. The View from all points was Splendid the West ward View, North, and North West a few miles distant was the "Blue Ridge Mountins," But still nearer was the Rapidan River 3 miles distant, the nearest ford being "Willis' Mills," whare I used to stand and wonder at the River Damed over, and the great old moss covered wheel slowly revolving and throwing the water off in beautiful showers.

And Orange Court House, where I was carried once behind the family carriage, along with the white children to see a Circus. and in the great crowd during a violent thunder storm I was lost and the carriage arrived at home without me and Mother and all, thought I had fallen asleep in my seat behind the carriage and been washed off in the "Mountain Run," which was very high. and had to be forded by the horses.

I wanderd about the Street at the Court House. (there was but one street) until dark and I had began to cry and wonder where I should sleep for I did not know any one in the village and could not think what I should do. and the crowds of country people were leaving the place very rapidly.

When all at once my godmother found me and soon had me safely packed away amoung a lot of Comforts and counterpains[1] in an old lumbering ox cart going to the next farm to my home. Where I arrived safe at about sunrise next morning. amidst great rejoicing.

A man was just getting ready to come and look for me. I was never allowed to go to another Circus at Orange Court House.

Once in a Month there used to meetings held for Divine Services (on the 4th Sundays I think) at "Mount Pisgah"[2] meeting house situated on the East end of the farm where Two Roads Intersected. I know they were "Baptists" because they used to Baptize in the creek close by.

To these meetings mother hardly ever failed to attend and take me with her, the church a large frame building with gallerys around for colord people to sit in. Tall pine trees surrounded the building and the horses and mules used to be fastened during the servicises.

Cakes Candy and fruits used to be sold there under those great pine trees on Sundays; which to my Eyes was always a great treat. I loved very much to go to the Meeting House as it was called then, because I never failed to come home with a load of cakes, candy and .c.[3]

These are my first recolections of attending a church and to-day in my 35th Year the Memory is bright as Events of to-day.[4]

Minute events of my reccolection will not likely interest you, so I must pass on with a glance at some of the most important events. ——

Very early in my infantcy other taught me that memorable little childs prayer not yet forgotten.

"Now I lay me down to sleep I pray the lord by soul to Keep".[5]

and then the "Lords Prayer" when I got older.

The usual routine of farm work went on for a long time (it seemed to me)

The slaves were treated kindly, and Every Sunday morning the weeks rations were issued to them from the great meat house.

Harvest time a festival of pies and meats fruits and vegitables would be set out in the yard on a great table in the shade. and the reapers, and binders[6] men and women seemed so happy merry and free, for whiskey was not withheld by the "Boss".

And then hog killing time (near christmas) when great fires were kindled and large stones made red hot, then placed into great hogsheads[7] of water until it boiled, for scalding the hogs. and Every body Was bussy, noisey and merry. Every one of the Slaves were permitted to raise their own hogs. and fowls and had a garden of their own from the Eldest man to me—

often at night. Singing and dancing, prayer meeting or corn Shucking.

"A Corn Shucking" is always a most lively time among the slaves. they would come for miles around to Join in singing and shouting and yelling as only a Negro can yell, for a good supply of Bad Whiskey and corn Bread and Bacon and cabbage.

At Christmas time the slaves were furnished with their new cloths Hats or Caps Boots and Shoes.

From the oldest to the little children they would be summoned to the "Great House" as they called it (the owners) and each man and woman would receive their christmas gifts namely Flour, Sugar, Whiskey, Molases e. c. according to the number in the family and they would go to their Cabins and for the next six days have a holiday and make thing lively with Egg-nogg, Opossum, Rabbits and Coons and Everything.

CHAPTER 2

Slavery

At about 4 years of age Mother learned me the afphabet from the "New York Primer,"[8] I was kept at my lessons an hour or Two each night by my mother.

My first great sorrow was caused by seeing one morning, a number of the "Plantation Hands," formed into line, with little Bundles straped to their backs, men, women, and children. and all marched off to be Sold South away from all that was near and dear to them, Parents, Wives, husbands and children; all Seporated one from another; perhaps never to meet again on Earth.[9]

I shall never forget the weeping that morning amounq those that were left behind, each one Expecting to go next.

It was not long before all on that farm was doomed to the same fate. and those that did not belong to the "Planter" had to be sent home to their owner.

The farm; and farming implements, stock and every thing was sold. and Mr Brown removed to Western Virginia.

Mother with me; and four other* children was sent to Fredericksburg Va.

where we all arrived safe after 2 days travel in an old "Road Wagon," soon thereafter my mother was sent to live to herself. that is to earn her own, and four little children, & living, without any help from our owners (Except) Doctors Bills.

* Louisa, Laura, Georgianna, and Willie.[10]

Poor mother struggled hard late and early to get a poor pittance for the children all of which was too small to help her. I was kept at the house of the "Old Mistress," all day to run erands and wait on the table or any thing else that I could do. At this time (the fall of 1848) Mrs. Taliaferro our old mistress, she haveing married a Mr. Frank Taliaferro who had since died, boarded at the "Farmers Bank," N.W. Cor. of George and Princess Ann Streets. Mother lived in a little house on George St. Betw. Sophia St. and the River.

I was dressed every morning (Except) Sunday in a neat white Apron and clean Jacket and Pants and sent up to the Bank to see what mistress might want me to do. possiablely she would have nothing at al for me to do. and if so, I would be ordered to sit down on a footstool, in her room for hours at a time when other children of my age would be out at play.

On Sundays I was dressed in my Sunday cloths (without the Apron) and sent to the "Baptist Church", cor Hanover and Sophia Sts of which the old Mistress was a strict member.

I used to have to sit where the old Lady could see me. As proff that I was there atal. At that time the White and Colord occupied the same church. only all the slave colord sat in the gallery on each side and the Free Colored sat in the gallery fronting the Pulpit.[11]

In the Afternoon of Sundays the Colored people used to have meetings in the Basement of the church. To which I was regularly sent and and Ordered, to bring home the text in order that the old Mistress might know whether I had been there or not.

Now the result of all this compulsinary, church attending was just the reverse of what was desired viz: I became a thourogh hater of this church and consequently, I resorted to all kinds of subtifuge imaginable to stay away from church.

I would go to the church door and then wait, until the minister would announce his text then commit it to memory. So I could tell when I went home, this no sooner done than I would hasten off, to the river to play with some boat or other, which I could always get or to swim or play marbles or, any thing in preference to sitting in church.

I soon became a confirmed Liar; on account of being compelled to go to church Too Much, that one church, I was scarcely ever allowed to go elsewhere—

CHAPTER 3.

Left Alone.

I had now arrived at the age of between 11 and 12 years, and had began to see some of the many trials of slavery.

Mother lived alone and maintained us, children, for about 2 years, perhaps. When Mrs Taliaferro came to the conclusion that mother, with my sisters, Lousia, Laura, Georgianna, and brother Willie would have to be sent to Staunton Virginia, to be hired to one R.H. Phillips.[12]

Accordingly they were, all fitted out with new dresses, shoes and Bonnett e. c. with mothers Bedcloths and some other few articles and then was in readiness for their long Jurney across the Blue Ridge Mountins in the month of December 1850 about christmas.

The Night before Mother left me (as I was to be kept in hand by the old mistress for especial use) she, mother, came up to my little room I slept in the "White peoples house," and laid down on my bed by me and begged me for her own sake, try and be a good boy, say my prayers every night, remember all she had tried to teach me and always think of her.

her tears mingled with mine amid kisses and heart felt sorrow She tucked the Bed cloths around me, and bade me good night.

Bitter pangs filled my heart and thought I would rather die, on the Morrow Mother and Sisters and Brother all would leave me alone in this wide world to battle with temptation trials and hardship.

Who then could I complain to When I was peresecuted? Who then would come early the cold Winter mornings and call me up and help me do my hard tasks? Whose hand (patting) me upon the head would sooth my early trials.

Then and there my hatred was kindled secretly against my oppressors, and I promised myself If ever I got an opportunity I would run away from these devilish slave holders—The morrow came and with tears and Lementations cars left with all that was near and dear to me on Earth[13]

A Week afterwards I heard they had all arrived safe in Staunton.

We wrote often to each other as circumstances would admit. of course, the white people had to write and read all the letters that passed between us. About this time I began seriously to feel the need of learn to write for myself. I took advantage of every opportunity to improve in spelling. I had to attend to cleaning Mr. William Wares. Room and he kept a large quanity of Books on hand amoung them "Harpers New Monthly Magazine,"[14]

I used to take much pleasure in reading (but imperfectly) short stories, which soon induced me to look for the Book with lively interest each month.

Two young men (white) used to sleep with Mr. Wm Ware of nights, named Roberts and they assisted me very much in spelling only.

For it positively forbidden by law to teach a Negro to Write. So I had to fall back upon my own resources.

CHAPTER 4.

Learning to Write

My Uncle George,[15] Mother's Brother was one day in the lot where I lived with Mr Ware and noticed me trying to copy the writing Alphabet as shown in "Cowleys Spelling Book,"[16] of that time, prinicipally used by those trying to learn to read or write. So he asked me what I was trying to do. I replyed, I am trying to write, see here and seizing the pen he wrote the following lines, on a peice of wallpaper

"My Dear Mother,

I Take this opporteunity to write you a few lines to let you know that I am well," ——

Now said he when you can do that much you ~~can do that much~~ you can write to your mother. He was at best a poor

writer, but the copy that he had Just given me was as good as
the best penmanship would have been because I could not get a
teacher of any kind, or a Copy Book that I could understand.
However I availed myself of the first chance I had to buy a 12
cent copy book. which was a most wretched concern, and with
its help I was most successful in laying the foundation of a very
bad writer, for there was nothing like form or system about the
thing. About this time I by some means or other, attracted the
attention of the Rev. Wm. I. Walker,[17] who was one day
hanging paper in the house where I lived.

And seeing my efforts at writing he kindly stoped, and
wrote me a very good copy of the Alphabet from which I soon
learned to write some kind of an inteligable hand and am still
trying to improve—But having never had a regular course of
spelling taught me. I am in consequence very defficent in every
branch of a common Education. So those who may be tempted
to read thees pages may possiably learn for the first time the
disadvantages of of slavery. With some of its attending evils.[18]

CHAPTER 5.

Sunday School. Visit to Staunton &c.

The Episcopal Church in Fredericksburq is situated, on the
North East Corner of Princess Ann and George Street.
Surrounded on the North and East by the grave yard; Fronting
on Princess Ann Street about Midway the square was a small
one story brick. In which I used to to go to a Sunday school,

Sundays afternoon and was taught the Cathacism and verses of
the bible was read to us to get by heart.

I do not think much good resulted from this school, for we
was not permitted to learn the A.B.C.'s or to spell, but Mrs
Taliaferro was most zealous in sending me to just such places
on Sundays as she would by this means know where I was by
asking Miss Olive Hanson, my Teacher, by the way she was a
most kind and gentle Lady and I often now think of her sweet
face and blue Eyes and feel a spark of gratitude for the efforts
on her part, for I really know she would have learned me to read
and wright if the laws had permitted her so to do.

Notwithstanding such stringent rules as there was laid down
for me on Sundays I resorted to lieing and deception in order to
get a few hours play that was not allowed to me during the
week. Often I would steal some body's boat and and with a lot
of a bad boys as I could find go up or down the river for a row
instead of going to to church where I was sent.

I had the greatest love for the river and boats,[19] and such
risks as I passed through them for fun. I would not now
undertake for any price. On Sunday in the early part of June or
July 1852 I was ordered to church (in the afternoon, as usual) and
instead of going. I met a party of Boys on my way to church,
and we soon made an agreement to go across the River, and to
"Coalters Fishing Shore" which was a nice secluded place for
bathing. So we went and was soon into the river in great glee.

But while we were all some what afread that the Oberseer,
or some one would drive us off—one of the Boys cried out here
come the Overseer!

All of us hastened out of the water to get our cloths and hide
in the bushes to get them on. I unfortunately ran in some vines

of "Pison Oaks."[20] We remained hid long enough to see there was no one after us. When after playing amounq the ~~Revenes~~ and rivelets wild flowers and Black berrys till near sundown we went to our homes, most of us with a lie in our mouths, all passed off well until the latter part of the following week: when I broke out all over with "Pison Oak." Mrs Taliferro nor any one in the house knew positively that it was the Pison Oak as they had not the least idea that I had been near any such Pison or Even in the cuntry any where atal. But they supposed I might have gotten hold of a peice in putting in wood a few days before.

Of course I told them I had not been in the country any where. The Doctor told Mrs Taliferro I had the humor in the Blood, and, after a due course of "Castor Oil". "Epsom Salts" &c &c. advised that I be sent out in the country for a few weeks, in order to save my life—And as the old lady was awful afread of sickness and the Doctors Bill, she concluded I should be sent to Staunton, Virginia and allowed to stay about 4 Weeks. I was delighted with the proposition and for fear she should change he Mind, I very <u>conveinently</u> began to get sicker than ever.

But hearing her remark one day that she thought I was too sick to travel, I made haste to be almost well the next day.

In due time, my cloths were made ready and packed in a letter valese with the following inscription on a Brass plate on the end Redmon, U.S.N. that was the name of the former owner and in charge of Mr. Phillips and his family. one night I bid farewell to my friends and was soon whirling over the R.F.P. Railroad on the way to Staunton, Va.

Arriving in Staunton the Second day after leaving home; the

meeting between Mother and sisters and Brother was a most
happy one and long remembered.

I had traveled from Fredericksburg to Hanover Junction by
the "Richmond Fredericksburg and Potomac RailRoad," and
thence to Woodville: by the "Virginia* Central Rail Road",
where we took stages and continued the jurnay across the Blue
Ridge Mountin, arriving in Staunton about 11 o'clock P.M. on
the second night from home.

I Remained in Staunton about 2 or 3 months, where I really
enjoyed myself visiting the Mountins and many other
interesting places.

The "Deaf and Dumb Institution," "State Insane Asylum,"
"Virginia Female Institute," and all combined to teach me the
same sad lesson viz: the white man's power and oppression of
the Colord Slave.[21]

In October Mr Phillips one day, told me I would have to get
ready and go back home to Fredericksburg that week by the first
opporteunity(!)

Now the opporteunity that they so much needed was, that,
some white person should take charge of my body and see to its
safe delivery in Fredericksburg. To be sure they might write me
a pass and put me in charge of a poor white Stage Driver but
he could only take me about 40 miles on my Jurney; where I
would have to be transferred to the cars: and in fact had to be
transferred so often that there was a fare possibility that I might
make my Escape and get to some free state! of course provided
I would do such a disgraceful act:

* Now called the Chesapeake & Ohio R.R."

However, within a few days from the time I was notified a "white man," one Dr. Dowling was duly charged with the responsibility of "seeing me <u>safe</u> home."

We left Staunton one afternoon, after a sad and affecting parting with mother and sisters and Brother. My ~~hart~~ heart was full and my voice choked with emotions and mother and children wept, as only those do, who do not know that they may ever meet again on Earth. Indeed either one might be sold on the Auction Block next day. The afternoon we left Staunton about 2 or 3 o'clock was a chilly dull looking day so frequent in the autumn, we crossed the "Blue Ridge", about dusk in a dense fog so thick that the horses attached to the stage could not be seen from the windows. We stopped at the little Village of Cotville for the night. That is till 2 o clock A.M. When we got up and resumed our jurney to Woodville and about 7 o clock that morning was snugley seated in the Eastward bound train.

An Accident occured on the Train about 3 miles west of Charlottsville to one of the Colored men employed on the train he was walking on a plank outside of the mail car, ~~which was~~

The plank about 10 inches wide ran length way the car from one end to the other. So by holding on to an iron Rod above. one might pass to and fro without going through the mail car. this man whos name was Scott was passing to the front of the car when he steped upon an Orange peeling or something or in passing by a fence that projected too near the cars he was dashed to the ground, Violently and his skull fractured. Some of the passengers seeing the accident Informed the Conductor who had the Train backed and taken the man into the car in which I was seated (the Niggers Car) the blood flowed freely from the wounded man's head and ~~years~~ ears until the Train arrived at

Charlottesville where he was removed and Died afterwards, I
heard.[22]

Another Accident;

The Locomotive broke down just as we stoped in at
Charlottesville for breakfast. We were called in to breakfast but
I could not eat any thing after seeing so much blood. Meantime
A Hand Car had been sent to the Next station, Dinwiddie, for a
Freight locomotive, awaiting there for our Train pass, After we
had been detained several hours and was just in the act of startin
on our jurney again with the boken Engine repaired a little by
the Engineer,

The Freaght Locomotive have in sight and after some
disput between the Engineers in reference to who should
take the Train, Moseby our first Engineer Claimed that
our Engine, ("Blue Ridge") would probably draw us to
Gordonsville quicker than the old Freight Engine, to which
that Engineer replyed if the Blue Ridge was allowed to start
with The Mail Train and should breake down he would not
render any assistance, again:

So it was decided that the Freaght Engineer and
locomotive, should take the Train to Gordonsville and Moseby,
with the "Blue Ridge," should follow and take the Freight from
Dinwiddie to Gordonsville.——

When our train arrived at Dinwiddie we could do nothing
of the Crippled locomotive behind us, but just before we arrived
at the next station; we had a good View of the "Blue Ridge",
and freaght train comeing at fine speed after us until we reached
Gordonville much behind time, and too late to stop for Dinner.
The Hotel Propitor sent in to the train by waiter Enough of
ham sandwiches for all the passengers, Gratis.

We then changed Engines for a beautiful one the "Rock Fish" at Gordonsville and I never Seen it since. The "Rock Fish," made splendid time to "Hanover Junction," where I changed cars for Fredericksburg. Those cars did not arrive at the junction until about 8.Oclock P.M.

Consequently I had several hours to look around about the junction. There was nothing of interest worth noteinq here.—It was about 7 years before I seen it again.

About 8. Oclock that night the Train From Richmond stoped and I was soon seated into a dark car with a lot of empty Mail Bags and Boxes around me with no living soul except myself—I was not doomed to solitude long, however, the cars had been running but a short time when one of the agents with a lantern in hand came into the car.

Exclaimed hilow boy! What are you doing in here in the Dark? I told him I thought this was the colord peoples car; where is your Master I soon told him who I was traveling with. Well come along with me, said he and led the way intos another car which I found well lighted, comfortable, quite full of white passengers, mostly asleep. I also soon fell asleep and was awake up in Fredericksburg.

The white people seemed very glad to see me probabley being releived from axiety of my possiable escap to Freedom!

CHAPTER 6.

The years of 1853–54. was passed in the usual routine of slave life. with its many sorrows and fears and fiting hopes of Escape to Freedom. So far as I was concerned I was kept unusually close, never permitted to pass the limits of the lot; after sundown without a permission and limited time to return, which must be punctually obeyed if I had any desire to go out again in a reasonable time.

On Sundays the same of "Rules" mentioned in Chapter 5. was strictly Enforced, which if disobeyed, at any one time would be sufficient cause to keep me in many Sundays thereafter Imagine a boy about 16 or 17 years of age in good health with many rolickinq fun loving companions playing in full sight of the house. on bright Sunday morning in the months of May or June, with a beautiful surronding country spread out for miles around visible to the naked Eye: With the sweet scent of clover locust Blossums, Hunnysuckle, Apple, Cherry, and various Fruit Trees almost Ripened, and all nature clothed with beauty, that can not be describe.—And that boy only permitted to see all this from an open window.

Not permitted to go out and see and smell the work of Him, who created all things. Imagine such a case. I say and you will have a very faint glimmer of my case at that time.[23]

I was very seldom allowed to visit any partys of young company except Fairs which was held for the benefit of one of the Colord Churches as there was not but Two.

The "African Baptist," and "Little Wesley," Methodist. Some time a fair would be opened at a privite house for the benefit of some poor person trying to make up money to finish paying for them selves or otherwise afflicted.

But nothing of the kind could be had without a permitt from the Mayor of the Town. Such specifying the time the fair should be closed. Which had to be strictly adheared to. All such

"Fairs" had to be held During the "Easter" "Whitsentide"[24] or "Christmas Hollidays".

To these fairs some times I was permitted to go which to me was almost a heavenly boon. It was at one of those fairs I frequently met Miss Annie E Gordon, having received an Introduction through a friend in a rather singular manner.

It was as follows, I hand wrote a Valentine a few months previous, for A friend of mine. Austin Bunday who had it Directed to this young lady.

I then obtained a promise from him that I would accompany him on his next visit to this lady and be introduced under the assumed name of Mr John Bunday his brother. this plan was duly carried out and Miss Annie Gordon was fully satisfied that it was my true name—until the Hollidays of "Easter," or Whitsentide 1853.[25] When by an accident, she over heard some other young ladies of my aquaintance call me by my right name. When she asked me what it meant, and I then told her of the fact: she seemed some what annoyed, but it was alright in the course of time. I had conceived a particular fancy for this young girl at first sight.

I was then very bashful and backward in speech. With probablely no kind of Idea of love making. Only this girl had a sad but very pretty face, and shy half scared look as if she thought I would bite her. She talked but little in my presence and then so low, you might have thought she was talking to herself. So I some times visited her at her Mothers or saw her at the Churches and in the street. I did not visit her often for a long time afterward ——

The African Baptist Church was situated on the N.E. Cor. of Sophia and Hanover Sts. (At the time I am speaking of the

"White people," worshiped in the upper part of the Building and the "Colord people," in the Basement.)

In the Spring of 1855. A great Revival of Religion prevailed amoung both White and Colord people of that church and a great many was added to its membership. Amoung those that joined at that time was manny young men and young women of my particular friends.

It was during this revival that I was Sincerely trubbled about the Salvation of my Soul. and about the 25th of May I was converted and found the Saviour precious to my soul, and heavenly joyes manefested, and began to be felt at that time, are still like burning coals; fanned by the breeze, (after a lapse of nearly 17 years) and Is to this day the most precicous assurance of my life, God grant the more faith and a better understanding, for these things let rocksand hills their lasting silance breake; And all harmonous human Tongues their Saviours praises Speak.[26] I was Baptized in the Rappahannock River at Fredericksburg, Va. by Rev. Wm F. Broaddus June 13th 1856. And many happy moments have I spent with the Church in its joys and sorrows. at that place. I was permitted to attend divine service on Sundays but at nights I was not allowed to go out but little—During my close imprisonment[27] (I do not know what else to call it) The "Word of God," was to me a source of unfailing pleasure. I became a close reader of the Bibl And Wrote many comments on different chapters which has since been lost.

It was during the autumn of 1856, that I Experienced my first attack of sickness of any duration.

Which soon developed itself into a severe and protracted case of Typhoid Fever which wore itself off in "Chills and Fevers," about 3 months from the commencement.

Be it said to their credit that Mrs Taliferro and her son Wm Ware. was the most attentive to me during my whole Sickness. I could not have been better attended to by my nearest relations—

I have often since wished it had pleased "Devine Providence" to have taken me from this world of Sin then—when I had not, as now seen so much of the Exceeding "Sinfulness of Sin"——

I Remained on the lot in service with Mr Ware and his mother until January 1st 1869.[28] When I was hired to Wm T. Heart;[29] next door, where I had to Drive Horses, Attend a cow help in the garden and everything else like work. But to me the change was very agreeable indeed all, Sunday and night restrictions was removed Except what was really nessesary. My clothing was abundant and good. My opportunitys to make money for myself was increased tenfold. I lived with Mrs Heart one year 1859.—

January 1st 1860 I went to live with Mssrs Arexander & Gibbs. Tobbacco Manufactorers. Where I, in a month or Two learned the art of preparing Tobacco for the mill. We were all "Tasked"[30] to Twist from 66 1/8 to 100 lbs per day. all the work we could do over the task we got paid for which was our own money, not our masters in this way some of us could make $3 or 4 Extra in a week—

The Factory weeks began on Saturdays and Ended on Fridays, when the Books were posted and all the men that had over work were paid promply on Saturday.

But if any one failed to have completed his tasks the lash would be generally resorted to—In a Tobacco Factory. the "Twisters" generally have one or Two boys, sometime women

for steming the Tobacco to be "Twisted," The Factorys is kept
very clean and warmed in winter from early morning till late at
night could be heard The noise of the macheinry and singing of
the hands in one incessant din—in a Tobacco Factory some of
the finest singing known to the colord Race could frequently be
heard—I was only permmitted to live one year in consequience
of the threatning position of the Southern States, the firm of
Alexander & Gibbs suspended operation. This year in the
Factory was to me more like "Freedom," then any I had known
since I was a very small boy. We began Work at 7. O clock in
the morning Stoped from 1 to 2 o clock for dinner—Stoped
work at 6 P.M. If we chose to make Extra work We began at
any hour before 7 and worked some times till 9. P.M.

the Sesession of South Carolina, and the the threatened
close of business between the North and South caused the
suspension of work in this factory Early in December 1860.

CHAPTER 7.

January 1st 1861. I was sent to Richmond, Va to be hired out. I
had long desired to go to Richmond. I had been told by my
friends it was a good place to make money for myself and I
wanted to go there.

So with a great many of old Friends I was placed in the
Care of Mr Hay Hoomes, hireing agent, and (on the cars)
started to Richmond where we arrived about 3 o,clock the same
day, and I was hired to one Zetelle, An Eating Saloon keeper
there was no liquors kept there.

I lived with him six months when he sold the place to a
man named Wendlinger, both of these men were low, mean,
and coarse. they treated their servants cruelly often whipping
them their selves or sending them to the slave jail to be whipped
where it was done fearfully for 50 cents.[31]

I got along unusually well with both men Especially the
latter.

I was living there when the Southern Slave holders in open
Rebellion fired on Fort Sumter, little did they then think, that
they were Fireing the Death-knell of Slavery, and little did I
think that my deliverence was so near at hand.

The fireing on Fort Sumter occurred April 12, 1861, and
from that time forward Richmond became the seat of the
Rebellion. Thousands of troops was sent to Richmond from
all parts of the south on their way to Washington, as they said.
and so many troops of all discription was landed there that it
appeard to be an impossiability, to us, colord people, that they
could ever by conquord.

In July 1861, the 21st day the Union Army, and the Rebels
met at Bulls Run and a great Battle was fought and the union
army was defeated. Already the slaves had been Escapeing into
the union armys lines and many thereby getting of to the Free
States. I could read the papers and Eagerly watched them for
tidings of the war which had began in earnest. almost every day
brought news of Battles. The Union troops was called "Yankees
and the Southern "Rebs", It had now become a well known fact
that slaves was daily making their Escape into the union lines.
So at Christmas 1861 I left Richmond, having been provided
with a pass and fare to Fredericksburg Va

I bid Mr Wendlinger and my fellow servants good-by They Expected me back the 1st of January again to live with them another year.

Soon after I arrived in Fredericksburg I sought and obtained a home for the year of 1862. at the "Shakespear House", Part of the time as "Steward," and the balance as Bar-keeper—My Master was not ~~much~~ pleased when he heard of my intention to remain in Fredericksburg that year; he seemed to think I wanted to remain too near the "Yankees", though he did not tell me these words.

The war was getting hoter Every day and the Yankees had approched within a few miles of the Town more than once. The later part of February 1862. the Rebs began to withdraw their forces from the Aquia Creek Landing which was then the terminus of the "Richmond Fredericksburg and Potomac Railroad", Early in March the Rebs began to fall back from the Potomack River;

The Town was now filled with Rebel Soilders, and their outrages and dastardly acts toward the colord people can not be told. It became dangerous to be out atal of nights.

The whites was hastening their slaves off to safeer places of refuge.

A great many slave men were sent into the Rebel Army as Drivers, Cooks, Hostlers, and any thing Else they could do.

The Firm of Payton & Mazine[32] who hired me were both officers in the Rebel Army. the first Captian in the 30th Regiment of Virginia the later Was a Lieutanant in the same Regiment, was at home, on the sick list and ~~was~~ in charge of the Hotel.

About the last of March there was a good deal of talk about Evacuating Fredericksburg. Which was soon after, commenced. by the 15th of April. Most of the troops had been withdrawn. On the Night of the 15th or 16th, the Yankees advanced and had a Skirmish, and drove in the Rebel pickets with some of them wounded and the others most frightfully scared.

The Propietors of the Shakespear now told me the house would have to be closed very soon in consequence of the near approach of the Yankees. and that I would have to go to Saulsbury, North Carolina[33] to wait on Capt. Payton the balance of the year.

I could not very well make any objections as the Firm had always treated me well and paid me besides, for attending the Bar for them, when I was hired only for a Dining room servant.

I was easily induced to change from the Dining room for $37.00 and Extra money every week.

So When I was told that I would have to go to Saulsbery I became greatly alarmed and began to fear that the object in sending me down there, was to be done to get me out of the reach of the Yankees. and I secretly resolved not to go But I made them believe I was most anxious to go.

In fact I made them believe I was tereblely afred of the Yankees, any way.

My Master was well satisfied at my appearant disposition and told me I was quite Right, for if the Yankees were to catch me they would send me to Cuba or cut my hands off or otherwise maltreat me. I of course pretended to beleive all they said but knew they were lieing all the while. As soon as they told me When I had to start, I Intended to conceale myself

and wait the approach of the Yankees and when once in the
lines I intended to go to Detroit, Michigan where I had an
uncle living.

CHAPTER 8.

April 18th 1862. Was "Good-Friday". the Day was a mild
pleasant one with the sun shining brightly, and everything
unusally quite, the Hotel Was crowed with boarders who was
seated at breakfast a rumor had been circulated amoung them
that the Yankees was advancing but nobody seemed to beleive it,
until Every body was startled by several reports of cannon.

Then in an instant all was wild confusion as a calvary man
dashed into the Dining Room and said "The Yankees is in
Falmouth." Every body was on their feet at once, no body
finished but some ran to their rooms to get a few things,
Officers and soldiers hurried to their Quarters Every where was
hurried words and hasty foot steps.

Mr Mazene who had hurried to his room now came runing
back called me out in the Hall and thrust a roll of Bank notes in
my hand and hurriedly told me to pay off all the servants, and
shut up the house and take charge of every thing.

"If the Yankees catch me they will kill me so I can't stay
here," "said he", and was off at full speed like the wind. In less
time than it takes me to write these lines, every white man was
out the house. Every man servant[34] was out on the house top
looking over the River at the Yankees for their glistening
bayonets could Easily be seen I could not begin to Express my

new born hopes for I felt already like I was certain of My
freedom Now.

By this time the Two Bridges crossing the River here was
on fire The match having been applied by the retreating rebels.
18 Vessels and 2 steamers at the wharf was all burnt In 2 hours
from the firing of the First gun. Every store in town was closed.
Every white man had run away or hid himself Every white
woman had shut themselves in doors. No one could be seen on
the streets but the colord people. and every one of them seemed
to be in the best of humors Every rebel soilder had left the town
and only a few of them hid in the woods west of the town
watching. The Yankees turned out to be the 1st Brigade of
"Kings Division", of McDowells Corpse, under Brigade Genl
Auger[35] having advanced as far as Falmouth they had Stoped on
Ficklins, Hill over looking the little town Genl Auger
discovered a rebel Artillery on the oppisite Side of the river
who, after setting fire to the Bridge was fireing at the Piers
trying to knock them down. the "Yankees" soon turned several
Peices loose on the Rebels who after a few shots beat a hasty
retreat; coming through Fredericksburg a a break neck speed as
if the "Yankees," was at their heels Instead of across the river
without a Ford, and all the Bridges burnt.

As soon as I had seen all things put to rights at the hotel
and the Doors closed and shutters put up, I call all the servants
in the Bar-Room and treated them all around plentifull and
after drinking "the Yankees", healths" I paid each one according
to Orders. I told them they could qo, just where they pleased
but be sure the "Yankees" have no trouble finding them.

I then put the keys into my pockets and proceeded to the
Bank where my old mistress lived who was hurridly packing her

silver-spoons to go out in the country to get away from the "Yankees". She asked me with tears in her Eyes what was I going to do. I replied I am going back to the Hotel now. After you get throug" said she", child you better come and go out in the country with me. So as to keep away from the Yankees. Yes madam "I replyed" I will come right back directly. I proceeded down to where Mrs. Mazene lived (the propietiors Wife) and delivered the keys to her.

Safe in the Lines.

After delivering the hotel keys to Mrs Mazene I then walked up Water St above Coalters Bridge where I noticed a large crowd of the people standing Eagerly gazeing across the river at a small group of officers and soilders who was now approaching the river side and immediately raised a flag of Truce and called out for some one to come over to them. A white man named James Turner, stepped into a small boat and went over to them. and after a few minutes returned with Capt. Wood of Harris' Light Calvary," of New York. Who as soon as he had landed proceeded up the hill to the crowd amoung which was the Mayor. "Common Council", and the Corporation Attorney; Thomas Barton.

Capt. Wood then in the name of Genl Auger commanding the U.S. Troops on the Falmouth Heights demanded the unconditional surrender of the Town. Old Lawer Barton was bitterly opposed to surrendering sayinq "the Confederacy had a plenty of Troops yet at their command" Then why did they

burn all the Bridges when we appeard on "Fickling Heights"?
demanded Capt Woods—Barton was silent. "The Orders are"
continued Capt. Wood "that if any further attempt is made to
burn cotton or any thing else, or if any Trains of cars Shall
approach or attempt to leave the town without permission of
Genl Auger the Town will be Immidialety fired upon.

The Mayor and "Common Council" hesitated no longer,
notwithstanding Lawer Barton's objection, and Capt Wood then
Informed the Mayor that he would be required to come over to
Genl Augers Headquarters the next morning at 10 o,clock and
sign the proper papers. He then bid them all good Evening and
having again Entered the little Boat he was soon rowed across
the River and in a few minutes thare after he was seen mounted
on horse back and being joined by scores of other Horsemen,
that had not been seen while he was on our side of the river.
Evedently having been concealed in the woods near by.

As soon as the Officer had left the Constables was told to
order the Negroes home which they did, but while we dispersed
from thereabouts a great many did not go home Just then.[36]
I hastened off in the direction of home and after making a
circuitous route I, in company with James Washington, my first
cousin and another free colord man left the town near the woolen
mills and proceeded up the road leading to Falmouth our object
being to get right oppisite the "Union Camp" and listen to the
great number of "Bands" then playing those tuchinq tunes, "the
Star Spangled Banner", "Red, White and Blue", &c.

We left the road just before we got to "Ficklin's Mill", and
walked down to the river. The long line of sentnels on the other
side doing duty colose to the water's Edge.

Very soon one, of a party of soilders, in a boat call out to the
crowd standing arround me do any of you want to come over—
Every body "said no," I hallowed out, "Yes I want to come over,"
"all right—Bully for you" was the response. and they was soon
over to our side. I greeted them gladly and stepped into their
Boat, as soon as James saw my determernation to go he joined
me and the other young man who had come along with us—

After we had landed on the other side, a large crowd of the
soilders off duty, gathered around us and asked all kinds of
questions in reference to the where abouts of the "Rebels" I had
stuffed my pockets full of rebel newspapers and, I distributed
them around as far as they would go greatly to the delight of the
men, and by this act won their good opinions right away. I told
them I was most happy to see them all that I had been looking
for them for a long time. Just here "one of them asked me I geuss
you ain,t a "Secessish," then, me "said I know why colord people
aint secessh, "Why you aint a colord man are you. "Said he," Yes
Sir I am "I replyed," and a slave all my life—All of them seemed
to utterly astonished. "do you want to be free inquired one" by all
means "I answered." "Where Is your Master?" said another: In
the Rebel Navy, "I said" well you don't belong to any body then.
"said several at once." The District of Columbia is free now.[37]
Emancapated 2 Days ago I did not know what to say for I was
dumb with joy and could only thank God and laugh.

They insisted upon my going up to their camp on the Hill,
and continued to ask all kind of questions about the "Rebs." I
was conducted all over their camp and shown Every thing that
could interest me most kind attention was shown me by a
Corporal in Company H 21st New York State Volenteers.

He shared his meals and his bed with me and seemed to pity me with all his manly heart. His name was "Charles Ladd,"[38] But our acquaintance was of short duration a few weeks thereafter the army advanced and had several skirmishes and I never seen him again.

It was near night before I thought of returning home (for though there was not as yet any of the "Union Troops" in Fredericksburg.) the Town was right under their guns and a close watch was being kept on the Town.

When my friends (the soilders) and me arrived at the River side we found the Boat drawn out of the water and all intercourse forbidden for the night. My cousin and his friend had recrossed early in the afternoon.

So I found I should have to remain with my new found friends for the night. However I was well aquainted in Falmouth and soon found the soft side of a wooden Bench; at Mrs Butlers[39] who had given us an entire room for the use of some soilders and 3 or 4 of us. A good fire was Kept burning all night in an old fashiond fire-place.

A most memorable night that was to me the soilders assured me that I was now a free man and had nothing to do but to stay with. They told me I could soon get a situation waiting on some of the officers, I had already been offerd one or two, and had determined to take one or the other, as soon as I could go over and get my cloths and Some $30.00 of my own.

Before morning I had began to fee like I had truly Escaped from the hand of the slaves master and with the help of God, I never would be a slave no more. I felt for the first time in my life that I could now claim Every cent that I should work for as my own. I began now to feel that life had a new joy awaiting

me. I might now go and come when I pleased So I wood remain
with the army until I got Enough money to travel further North
<u>This was the First Night</u> of my Freedom. It was good Friday
indeed the Best Friday I had ever seen Thank God—xxxx ——
we were all asstire [astir?] very early next morning for the
soilders had a sad duty to perform.

The night before they captured Falmouth, they, while
advancing suddenly in the darkness found the Road Barracaded,
and the Rebels concealed, close by who fired upon the
advancing troops where the Road way cut through a hill and
killed 7 and wounded several.

The Funeral was one of the most solomn and impressive I
had ever witnessed in my life before. Their company (Calvary)
was dismounted and drawn up in line, around the seven new
graves. Which had been dug side-by-side

The old Family Burrying ground wherein those New made
graves had been dug contained the Bones some of the Odest
and most wealthy of the Early settlers of Falmouth. On some
of the Tomb=Stones could be dimly traced The Birth-places of
some in England, Scotland and Wales as well as Irland. And
admidst grand old Tombs and Vaults, sorrunded by noble
Ceders through which the April wind seemed to moan low
dirges; There they was now about to deposit the remains of
(what the rebels was pleased to term) the low born "Yankee,"
side-by-side they rested those seven coffins on the edge of these
seven new made graves. While the Chaplins fervent pray was
wafted to the skies and after a Hymn (Windham)[40] had been
sung, those seven coffins was lowerd to their final resting places.

And amidst the sound of the Earth falling into those New
made graves The "Band of Harris' Light Calvary broke forth in

dear old "Pleyal's Hymn[41] and when those graves were finished
there was scarcely a dry eye present.

And with heavy hearts their companions left that little
Burrying ground some swearing to avenge their Deaths. A few
hours after the funeral. The Mayor of Fredericks accompanyed
by several of the leading men of the Town, crossed the River
and came to Headquarters, where, the Town was formelly
surrenderd. Troops was then sent over to take Possion of any
thing nesserary for the "Union Army and to guard the Town.
x x x x x About 3 weeks afterward I was Employed as the
Headmasters of Magr Genl Rufus King, who was then
commanding the 1st Division of "The Mc Dowells" Corps
"Army of the Rappahanock", as mess servant. I had left Genl
Augers Headquarters some Two weeks, and was staying with
my old friend John Walters, at the "Phillips Farm," about 1 1/2
mile distant Eastward from Fredericksburg. Genl King had
taken possession of the "Phillips House," some 10 days,
previous, and one Sunday morning while I was strolling about
the yard of the Headquarters looking at the officers and soilders,
when I was accosted by Captain Charles Wood Aid-de-Camp
to Genl King as follows. Well man where do you live? "I have
been living over there Sir." All my life. I answered. Who did you
belong to: Thomas R Ware Sir. Where is he now?

In the Rebel Navy on Board the Jamestown. "I replyed."
Ah—then you belong to me! "Said he laughing. Thank you sir!
"said I." Tuching my hat to which there was a general laugh
amoung them standing by. —— The Captain after telling me I
was free to do as I was pleased. Then Engaged me at $18.00 per
month to take charge of the general and his staff officers, mess
and keep things in order generally.

I was conducted right to the General,s kitchen where Every thing was placed in my charge (after an Introduction, to the cook. Whose name was Ransom Law: Detailed from the 6th Wisconsin Regiment and Cook) with an order to make some Beef Hash. Which the Officers, at Breakfast announced splendid. Much to my delight, for I had my fears of being able to please them. But I succeeded beyond my Expectations and was soon a great favorite with all, from the General to the Orderlies.

One day soon after I was Employed General King Sent for me and when I reported to him. "He said John go to the stable and tell Erastus (his hosteler) to give you my Horse Charley, Captain Wood wants you. I bowed and ~~hass~~ hastened to the stable where I Found "Charley" saddled and Brideled already. I mounted, and the Horse being a thourohh bred I was not a little frightened at his disposition to walk on his hind feet, instead of all four. But as soon as I could get him in hand and look about myself. I discovered Captain Wood with a Company of Calvary arriving a little distance off for me; to joine them. They then proceeded down the road toward Fredericksburg I was ordered to ride next to Captain Wood and Col Fairchilds, who was riding by the side of each other.

When we arrived at the "Old Ferry" opposite the Town a Bridge had been nearly completed across. built on old Canal boats[42] a few minutes delay and the last planks was laid. I now ascertained that I had been brought along to act as a guide in identifying the prominent Rebels of the Town, and after they had crossed the River and Entered the Town we proceeded directly to the Post-Office, then kept by one R. T. Thom. Capt Wood called me to point out Each place and to name each person required.[43]

Mr Thom was then placed under arrest. Capt Wood then taken me and left all the Officers, and said where is the "Sheakespeare House," this way sir. And we soon dismounted in front of the Hotel. He entered with me and gave Mrs Payton orders that nothing should be sent out of that house, Except on an Order from Genl King

From there Capt Wood rode to the Mayor of the Town and other prominent rebels. Some 25 or 30 in number and the next morning they were Escorted to Headquarters under arrest, and were sent to Washington City as prisinors **

CHAPTER 9.

Gen= King soon, after the incidents narrated in the proceeding chapter stationed his headquarters in Fredericksburg at my old house "The Farmers Bank." and as a natural consequence Every body and Every thing seemed to take new life. There were some few Rebel Simithythzers [sympathizers] amoung the Colord people But they kept very quite.

Hundreds of colord people obtained passes and free transportation to Washington and the North. And made their Escape to the Free States.

Day after day the slaves came into camp and every where that the "Stars and Stripes", waved they seemed to know freedom had dawned to the slave.

May 23rd "The Battle of Front Royal"[44] was fought and Genl King was ordered to march to "Catletts Station", on the Orange and Alexandria Railroad, about 38 miles from

Fredericksburg. Whence a part of ~~his~~ the Division was
transported by Railroad to Front Royal but in consequence of an
accident on the Railroad by which several soilders was killed and
wounded. The remaining Troops, with the General and his staff
and followers had to March overland to Hay Market, distant 17
miles – The servants horses had been sent on the last and there
being no Extra Horses John Brown (The Generals hostler) and
me and several others (colord) men had to foot it with the
soilders.

There was Several Regiments of Infantry and Calvary, along
with the Headquarters wagons and Tranports but all to heavy
laden to give <u>us</u> a ride for a few minutes.

Many of the soilders laughed at us and cherred to see us
dismounted as well as they. We left ~~eatells~~ "Catletts Station"
about 3 o,clock in the afternoon the weather was very pleasant
but cloudy. We were hurried acrofts fields and meadows and by-
paths along rugged roads and through the yards of Farm Houses.
While the terified women and children huddled together (white
and colord), as if for protection from the Invading foe.

About 10 Oclock that night we halted and camped for the
night. Major Coons was in command of a company of Harris
Light Calvary, and our Headquarters teams as soon as the soilders
pitched their tents. John Brown me and 4 of the Generals
Orderlies who was also dismounted and all of us without tents
began to look about for some shelters of some kind.

We had pass a little farm House about a mile back. And as
it had began to rain quite hard, some one proposed that we go
back and take shelter in one of the out houses. Which was
agreed to and we were soon searching all the buildings for some
occupant but we found the place entirely deserted.

We soon made a fire in a lower Room and proceeded to make coffe and then to make our supper off of "hard Tack", and some salt pork. While we sat partaking of our needed repast. We were suddenly alarmed by hearing a horse dash into the yard and some one in a loud voice demand what we were! and what were we doing here?

Seizing our ~~arms~~ pistols and sabres haistly and looking out the door, and found Major Coons and a squad of Calvary at his heels, he swearing "like a Trooper", ordered us into camp instantly and not to leave again on perial of Death.

We obeyed and followed into camp, which was dark and silent as the grave. The camp was a marsh wet spot covered with coarse grass near a Creek. The grass was too wet to lie on if it could be avoided — — after a Whispered consultation (no one was allowed to talk loud) We, again one by one glided silently out of camp skirting [?] the fields reached the little Farm House. After Having been told the "Gurrilas", had been there the night before and captured some of our men, and pulling an old Bedstead to pieces and with one or two fence rails secured the two Doors, we laid our knapsacks on the floor against each Door for pillows and our pistols under our heads and our sabres close by for immediate use in case of an attack, during the night We were soon in a sound slumber sweet and refreshing from which we did not awake until 6 oclock next morning.

One of the boys in the mean time had unfastened the Door and on looking over the Camp was not a little surprised to find Every man and beast gone. And worse than all the fallen rain had hid every trace of the direction the troops had taken.

Following the road northward was our only hope, and with our unusual good chance of being captured by "Mosby[45] and his group" of gurillas – Late in the after noon we came up with the rear of the army near HayMarket. Persuing the road onward we found Genl King and his staff at a Tavern in the Village.

The General had just given orders to the Headquarters teams to go on to Garnesville on the Orange and Alexandria R. R. Where we Encamped for a few days—during which it rained most of the time.

Early one morning, on the fourth day I think, we received marching orders to Warrenton. the road a turnpike was an unusal good – one for this location being well settled with stone over which the wagons ambulances and troops, moved with comparative ease the day was a beautiful one, and the road was dotted here and there with fine and stately old mansions.

Surround by growing wheat and corn fields, and every indication of for wealth and prosperity.

About 3,Oclock in the Afternoon we arrived in site of Warrenton and Entered the Town. General King at the head of his Division with Bands playing "The Star Spangled Banner" "Red White and Blue", &.c&.c——

Crowds of Rebels stood on the court-House steeps and looked vengance at us as we advanced in to the Town. General King, Established his Headquarters in the Town at the Warren Green Hotel. The Troops encamped out side of the Town on the heights—The scenery is beautiful around this town. Mountins Hills and Valleys being coverd with splendid vegetation this season of the year.—The "Fauquire White Sulpuure Springs"[46] is located only a few mils from this place, and was much frequented before the war by the wealthy in search of health and enjoyment during the Summer.

The officers and servants had very good rooms assigned to them during our stay here. The hotel furnishing all nesserary accomadations. Our stay was of short duration, however. We arrived on Wednesday and Sunday morning some of us went to a church on one of the main streets. where one of our chaplins was to preach.

The opening hymn had been sung a prayer and chapter read, when and orderly was seen to approach the pulpit with a letter or order in his hand — — which he handed to the chaplin and haistly retired.

After reading it the chaplin arose and said "Every man is hereby ordered, to report to his respective quarters immediately." That was enough! When the church was left—the first news that we herd was "the rebs is advancing" xx. In Two hours the camps was all broken up. The Headquarters evacuated and the Calvery advancing toward "Catlett Station" again.

It was about 3 O,clock P.M. when the General with his staff and followers left the Town amidst the prayers and good will of the colord people that remained behind.

Hundreds of colord men, women, and children followed us closely on foot. Poor mothers with their Babys at their breasts. Fathers with a few cloths in Bundles or larger children accompanying them followed close in the foot steps of the soilders. Seeming to think this would be their surrist way to freedom. The distance from Warrenton to Catletts Station was 12 miles and these poor souls would be permitted to go to Washington where they were provided for by the U.S. government as "contrabrands of war"47 So they would be sent

down free of cost by the Railroad.——— We camped just
below "Catletts Station" that night. And resuming our march
next morning camped at our old Headquarters opposite
Fredericksburg at the "Phillips House".

But a few days after we were ordered across to Fredericksburg
and made the Headquarters in the "Farmers Bank". My old
home again this afforded me a great pleasure of being back with
old Friends, my grand mother and aunt lived there and kept
things together for the old mistress hid away in the country.

I occupied my old room for the first since I had Escaped
and I surely was never so happy as then and probably will not
soon forget it. Soon—Genl McClellan was advancing on
Richmond from Fortress Monroe, about this time. And soon
the orders was given to advance from our Headquarters When
"Gibbons Brigade" was sent on as far as Milford Depot and
Bowling Green. Genl Kilpatrick was making things lively with
his calvary every now and then and capturing prisoners and
arms.

We were awaiting marching orders when one day, a dispatch
was sent haistly recalling the the advanced troops and ordering
the Evacuation of Fredericksburg and an Immediate advance to
Culpeper Court House.

CHAPTER 10.

Sunday morning August 10th 1862 dawned bright and warm,
and the indications of a warm day was apparent to every one,
the whole Division was moving around Fulmouth Va and

Calvary Artillary, Infintary, and wagons and Ambulances was filling the roads that led toward Culpeper County. Information had reached our army that Gen Banks had been attacked at or near "Ceder Mountain", in Culpeper County and we were hastening to reinforce him. Genl John A. Pope[48] was then in charge of the Army of Virginia. and his Headquarters was near "Banks, xxxxx.

On leaving the camp at Falmouth our Division march rapidly to "Ellis' Ford where they Forded the River, the Infintry devesting themselves of their pants and the water not being over waist deep there.

We found the ford guarded by the 106th New York Regiment. Our General stoped here from about SunSet till 2 Oclock at night. Some of our boys went to the Kitchen belonging to "Ellis" and tried to purchase some Hot Biscuit which the colord women were bakeing for sale but the Door was guarded by the New York men, who had orders not to let our men enter. So our Boys thought they would get even with them and lateer in the night some half dozen of them entered the mill and after stealing about 30 fine Hams turned the water on, and left the mill runing without any thing to grind. About 2 O,clock that night whispered orders was given for us to mount, and after crossing the river, day broke and found us miles away from "Ellis' Mill.[49]

As the sun arose above the Mountains the air loaded with the rich perfume of clover and wild flowers. And the heavy mountain dew looking like drops of silver on the rich leaves and blossoms.

We had ridden past regiment after regiment of our men and the General and his staff was almost out of sight a little a head,

there being a turn in the road just there. I was looking around
at the beautiful mountain scenery around me. Each side of the
road being thickly lined with low ceder and pine. When I was
suddenly startled by the report of a Rifle near by and the whistle
of a "Minnie" ball close to my head. I dove the spurs into my
horses' flanks and hurried forward to rejoin our men.

We were now nearing the Battle field, and we were in
momentary Expectation of an attack.

About 3 O,clock General King and his staff halted at
"Strasburg" for a rest and lunch

While resting here the General called me and sent me back
to the wagon train for some whiskey that was on hand put up in
pint flasks.

I rode back and found the train about 2 miles in the rear
awaiting orders to move onward. After obtaining the whiskey,
I hastened back to Strasburg and found the General and his
staff had just rode off toward Ceder Mountain. It was now
about 5.00 and following the direction they had gone I soon
ascertained they had left the main Road and crossted the fields
from Strausburg to Ceder Mountain (about 10 Miles) the roads
was packed Crowded and Jammed with Calvary, Artillary,
infintry wagons, Contrabrands, refugees and Cattle. I pressed
my horse on toward the front passing Brigades and regiments
until I had overtaken the Artillry which was in advance. Night
had now overtaken us now I Received orders not to advance any
further than the artillery as they were feeling their ways to avoid
a surprise by night. The Reble pickets had been driven in earlier
in the Evening—

Woods on both sides of the road here was densely thick and
we did not know what moment the rebels might fire on us.

We soon approached a part of the road that was fenced in
with a rough stone Wall of great thickness and about 4 or 4 1/2
feet high the order was passed (in whispers) to dismount and lye
down on the ground which was done and we remained in that
position for some time.

Finnally the order to mount came in whispers. A portion
of the wall had been removed in order to let the artillery and
wagons and calvary pass in several places. After advancing some
distance up a hill we came to a halt and camped for the night.
Soon after we got orders to water the horses and in squads, we
proceeded to a little stream not far off and waterd our horses.
I also filled my canteen with water and drank freely of it and
learned afterward that the spring from which it ran, had been
poisioned a few days before. The only effects I felt from it was
rather Violent pains. With a burning thirst. I fed my horse and
slept in an ammunition wagon that night up[upon] Boxes of
Bumb-Shells.

Early next morning I rode off in search of Genl King and
found him and his staff after a few minutes ride where he had
a kind of shelter made of green corn stalks from a field just
accross the road.

A Few hours after I arrived our Headquarters teams came
up and we pitched Into the hams sliced and by sticking a stick
through the slices soon broiled enough for a good meal by
holding it to the fire.

A few shells from the rebels soon put an end to our cooking
by the smoke from our fire serving them as a mask.

During this afternoon we received orders to "fall back", to
"Fleet's" farm Where The General established his headquarters

for the time being in close proximity to Gen John Pope's Headquarters. Gen Shields and Gen Banks Major Genl Pope being their command of the "Army of Virginia", great preparations were being made for some important move. Orders had just been Issued for the Discharge of all servants[50] Except hostlers and Cooks. Of course this did not affect my case.

But A Reward of $300.00 had been offerd for my head in Fredericksburg and knowing if I should be captured by the rebels I should be taken to Richmond Va where I was well known and no doubt be immediately hung or shot for being with the "Yankees."[51]

I Therefore obtained permission to Visit Fredericksburg going by way of Washington City.

Genl King Willingly game the desired permission now, as there was no fighting going on just then. The Hospitals at Culpeper Ctc.[Courthouse] was crowed with the Wounded. And the Dead had been buried from the last Battle field.

The Stars and Stripes waved proudly from the different headquarters of generals and Colonels. The music from the Bands echoed and reached across hill and, Vale as cheerfully and gay as if there was non missing from the last earthly roll-call.[52]

But I am digressing—Having obtained my pass and money I Exchanged my blue pants for a pair of different color and bidding farewell to my companions in camp, with a sad heart was soon on my way to Culpeper C. H. where I took the train for Washington and arrived about 2 o,clock that night. Next morning I proceeded to Genl Pope's Headquarters on 17th Street. opposite the War Department, and obtained a pass to Frederickburg and going to 7th Street Wharf took the Steamer

Keyport (then in government services) for Aquia Creek and on landing I found that there was no passengers permitted to go over to Fredericksburg til next day.

A locomotive and train of flat cars was waiting on the track to take over a lot of troops of Genl Burnsides[53] command and while they were getting on the train I got on with them without the knowledge of the proper officer of the load. And was soon after landed at Falmouth Station about one mile this side of the Rappahaneck River. I walked over to the Town where I found my wife, as well as might be expected; They were all greatly surprised to see me when they supposed me to be at Culpeper County with Genl King. I Remained at home about one week Enjoying my freedom with friends and acquaintances. The old rebel citizens showing evident marks of displeasure at my appearance amoung them. They regarding me in the light of a spy, or traitor to their cause. I had intended now to stay at home and make a living and after a while, perhaps, to go north some where. When my wife would possibly be able to go with me, as the movements of both Armys (Union and Reble) were quite uncertain I did not know what minute the present force under Genl A. E. Burnside might be ordered away their troops were continually moving to and fro and heavy fireing had been heard, for several days, in the direction of Culpeper C. H.

And a great many soilders had crossed from Fredericksburg to the north side of the river and disappeared soon after, which gave rise soon that they Yankee was falling back. My wife and friends advised me not to let the Yankees leave me Behind if they did fall back. All they firmly believed the rebels would take my life.

My wife's mother had not spoken to her or me since our marrage and she had forbidden Annie to darken her door way again[54] or I should have gone to Washington with my wife and settled down but as it was my wife was not in a condition to travel far without and elder female friend. So Mrs Jackson and others of our friends advised me to leave at the earliest opportunity for the sake of safety and when my wife should be able she could come to Washington.

CHAPTER II.

Sunday afternoon Aug 31st 1862 about 4 oclock P. M.

Our attention was called to a dark smoke over in Falmouth and going to the river shore. We discovered that the Union Troops Were burning their Bakerys, which was very extensive and breaking up their camps in haste.

With a sad heart I returned to tell my wife the bad news for I knew well what it meant that our troops were Evacuating Falmouth, I bid my wife good-by and hastened to the Headquarters of Col Kingsbury, Provost Marshall of the Town to ascertain the facts in the case. I found Col Kingsbury in his office and stepping up to him saluted him and said Colonel I heard sir that the Union troops are going to Evacuate the Town is it so sir. What is it to you sir. Go out of this office, "said he." In a stern commanding voice. Bringing his clenched fist down on the table in front of him—I beg your pardon Colonel, "I said" I am Genl King's mess-servant. well what are you doing

here I just came down on a Visit to my wife sir. And don't want
to be left behind if you all are going. Well (in a milder voice
now) if you want to get away go right accross the Bridge within
15 minutes

He gave me a pass and I was soon across the Wire Bridge. I
had not had time to go back home for any clothing or money
and I had only 50 cents in my pocketts—When I crossed the
bridge I noticed shavings and Tar placed at intervals on several
different piers with kegs of powder near by.

After crossing the Bridge I hastened to the Top of the Hill
at the East end of the Bridge and looked back at the town that
had given me birth with a sad heart and full eyes thought of
some of the joys I had felt within its limits—But now
compelled to fly from it for my life, for daring to make my
Escape to the Union army and with a price fixed upon my head
if caught. I could not help weeping, (though it was not manly)
as I looked back, and thought of my poor young wife, who
could not fly with me. The Rebels was at that very minnit
swarming the Heights, West of Fredericksburg. And I know
not, but they might take vengeance on her as I had Ecaped
they could lock her up in jail or any thing else and who would
protect her.[55]

My Solequay [soliloquy] was interrupted by a tremendous
Explosion that could be heard for miles around, and shook the
Earth like an Earthquake. The Flame shot upward hundreds of
feet into the air – and as suddenly all was silent as death. But
the Wire Bridge was gone to ruins and the rebels victorious
short rang out over the Heights of Fredericksburg,***

Between 9 and 10 oclock that night I laid down in and out
building at the Lacy Farm with the Union Army encamped all

around me. When I awoke next morning a little after day light
not a soilder could be seen any where about. The whole of
"Burnsides" Division had fallen back toward Aquia Creek on
the Potomac River 15 miles distant. My case was a critical one,
now indeed. A rain late in the night had completely hid the
track of the army but I soon struck out for the railroad and after
following it for some time I left it and persued the road toward
Bell-Plane Landing 9 miles distant; after walking for some time
I discovered the fresh tracks of Horses going the same way that
I was, but soon I could hear voices. I stoped and listened for I
did not know whether they were Rebel gorrillas or the Union
army. I hid in the thick undergrowth close by till I caught site of
a blue coat. When my heart gave a great leap for joy, I was soon
once more in the Union line, and about 2 Oclock arrived at
Aquia Creek Landing.

I found the soilders encamped for miles around and
Hundreds of steamers and Transports of all discriptions
awaiting to receive their living Cargos, which was being shipped
in all haste. When I got there the mail steamer Keyport was
nearly ready to leave for Washington D.C I had several old
passes certifying that I was a servant at Genl Kings
Headquarters (but his headquarters was now up near Culpper
Court House) and consequently his passes was not respected.
So the Sentinel guarding the gang plank from the wharf to the
Steamer forbid me going aboard without a pass from Genl
Burnside whose Headquarters was nearby. I stood around a
few minuits watching an opportunity when the Sentinel stood
reading another pass. I bounded across the gang plank and
concealed myself for a while until the Steamer got off from the
Wharf. I then came back and arrived safe at 6th Street Wharf

in Washington D.C. on the night of September 1st 1862 in a
hard rain.

My grandmother, Aunt[56] and her 4 Children all slept on
14th St that night and next morning walked to Georgetown
where we had friends My grandmother Aunt and the children
soon found some place to stay, and I obtained board at Mrs
Boons at $250 per week. My next object was to obtain work in
order that I might pay my board and get a change of clothing
for I was sadly in need of them. I had no trade then and knew
not what to do. But soon learned to turn my hand to most any
thing light. There was a plenty of heavy work. Such loading and
unloading vessels and steamers but that was mostly to heavy for
me as I was not very strong[57] but finally obtained a place
Bottling Liquor for Dodge &c at $1.25 per day which lasted for
some time.

Journal of
Wallace Turnage

◆▶◀◆

Wallace Turnage's apology for his book. My book is a sketch
of my life or adventures and persecutions which I went through
from 1860 to 1865. I do not mean to speak disparagingly of those
who sold me, nor of those who bought me. Though I seen a
hard time, it had an attendency to make a man of me.

I under took to write these few stages some time ago, but I
become tired and weary of the tediousness, and finally left off
writing. But when I thought some one might be pleased to hear
of my adventures I concluded to commence again and I have
succeeded in writing what I have. It is not all of the details of
my hard ships, but merely a sketch of that which I think would
be most interesting to those who shall approve of my book. I
could say more of the south and of its fertile soil, but I don't
think it is necessary. I will also beg my reader to excuse my
ungrammatical and desultory biography because my kind reader
can see that I have been deprived of an education, and what

knowledge I have to present this biography to you, I learnt
during that time and since I escapted the clutches of those who
held me in slavery.

In the year of 1860, I was carried to Richmond, V.A. and
sold to a man by the name of Hector Davis. Now I could not be
sold rightly until all of the people's children I belonged to were
of age, so the older one got married, so she was allowed to draw
her part. Though after she had drawned me, she was not to sell
me out of the family, for her Brother was my father. For all of
that she and her Husband made a plot with one Mr. Reuben
Wallace to take me to Richmond and sell me. So they told me
that they was going to take me to More head City to nurse[1] and
I could come to see my Mother when I wanted to. so I thought
that was very nice and I consented to go, thinking all of the
while, that I was going to More Head City to nurse. but I was
greatly mistaken. so instead of going to Morehead city, I was
carried to Richmond and was sold to Mr. Hector Davis.[2] He
gave nine hundred and fifty dollars for me. I was kept then as
his auction and office boy. so I lived in Richmond some time,
taking people from the jail to the dressing room and from the
dressing room to the auction room. Well at last there was a man
who came from the south, he made it his business to come once
in the Spring and once in the Fall to buy in his winter and
summer goods.

So he always came through Richmond, and bought him
a slave or two also. so Mr. Chalermers,[3] for that was the
Gentleman's name, saw me and said he would like to buy me. so
Mr. Chalermers was a very clever man to converse with so I
liked him splendidly. he asked me would I like to live with him,
I told him I would. then I asked how far did he live, he said that

he did not live far, so Mr. Chalermers give Mr. Davis one
Thousand Dollars for me.

But mind you I didn't know that Mr. Chalermers had
bought me. I went on doing my, work as usual until the evening
before I went away the next morning. so as I was going along
Franklin Street the evening previous to the morning I went
away, I heard old Mr. Lee say take that boy up-at once, so that
he can go away with the drove⁴ in the morning. so I was put in
jail. Now Mr. Lee kept a jail also, a very old man. so the next
morning I and all of the Rest was ushered out in time to take
the four Oclock train via Petersburg V.A. and from thence
southward. so we in two days and two nights and a half day we
arrived in Montgomery Ala. and from thence to Mobile Ala.
and from thence to Pickensville Ala. where the man that bought
me lived. Now Mr. Chalermers had bought another man in
Richmond by the name of Wily, so we was in company often on
but I did not know that he belonged to the same man that I did.
Howe[ve]r me and Wily was carried to Mr. Chalermers House,
so I saw that he was the same little man that talked with me in
Richmond. Mr. Chalermers was a very rich man, he had about
seventy five slaves.

Well he kept us up to his home house for a few days and
then sent us down to his plantation to be initiated according to
their way of treatment. so we arrived at the plantation in the
evening or about noon. so they sent me out with the other
hands to burn brush. Now these colered people was very glad to
see me to learn something from old Virginia, and in the height
of their conversation they left off all of a sudden to speak to me
and began to work like they was going to tear every thing to
pieces. so I casted my eye across the field and I saw a common

looking, shabbily dressed man coming with a bull whip in his
hand. so I found out this was the overseer. so he come up to
his hands and made one of the women pull up her dress and
whiped her shame ful. Then he shook his whip at me and said
hurry up young man. so I tell you I thought that was very rough
for the first day. So the day passed away and we all went home
and got our supper and those people was Fidleing and dancing
as though they had no oppression at all.

Well the night passed away. the next morning I heard the
horns blewing all over the neighborhood about Four O-clock, so
they had me up to take charge of my Breakfast and dinner for
the day; it went very odd to me, but I had to get use to it. Now
the overseer would make those women strip right before the
men and whip them under their cloths on their naked skin,
shamefully, and he would do likewise to the men. Now when
I saw such whipings put upon those people I thought that
they could not work; but I found that the poor place of
accommodation had harden them to such treatment that they
had got use to it. I then said to them, look here, I will not take
such whipings as I see the overseer give. I can't work after wards
said I. The men said to me son, what are you going to do, we
are men and we have to take it. I told them that I would run
away. they said where will you go to. I told them that I would
go in the woods and go back where I come from, but they said
you can never get back any more. However during that spring
and summer the Overseer whiped me three or four times, but
not very severe.

Well every thing passed off very well until the Fall of the
year; long about corn gathering time, He had us all gathering
corn. so this made our hands very sore. So much so that we

could not pick as much cotton and give the satisfaction that we
had been giving. so there was four including me to be whiped.

Well the overseer began to weigh the cotton, and when he
got around to them that didn't have cotton enough he told them
to stand back until he got through weighing cotton and he would
see what was the reason they could not pick more cotton. so
when he got around to me he told me to stand back with the rest
of them and he would see why I couldn't pick more cotton. Well
when he got through weighing cotton, he took his cowhide and
made one of the women lay down, and pulled her clothes over
her head and made the other woman hold her and her clothes
over her head. he give that woman about two hundred lashes and
I thought that was enough except he was going to kill her. I
could see the skin fly near about every lick he struck her.[5] then
he made the woman that held the one he had whiped lay down
and made the one he had whiped hold her clothes over her head.
Then I thought if that was the kind of whipings he gave them I
would not stay and take mine. So I saw my chance while he was
whiping to make my escape, so I left.

Now when he was through whiping this woman, he was
then ready for me. so he called for Wallace next, but no Wallace
was there. Then he said I bet that little rascal has run away. I
could hear him calling me, but I would not come back. So I was
then in the woods with the wild animals and only about fifteen
years of age. I suffered very much for some thing to eat,

Well I wandered a way off in the neighborhood, but I got a
very little to eat. Now this being my first runaway, I didn't know
where to go. However I wandered back in the neighborhood of
the Plantation looking for something to eat. Now my master
had told some of his people on the plantation if they seen me to

tell me to come home to him up to his home and he would not let the over seer whip me. so I went to the plantation, as I said, to get some thing to eat, and I chanced to find one of the men's doors open and he was gone out to run off some dogs, to keep them from awaking the overseer. so I went in of course, and found a pot of bacon and cabbage; but before I could get out the man came in and caught me there. but he know me and told me not to be afraid. he did not say any thing to me about eating up his victuals. So he told me that my master said if I would come home, he would not let the over seer whip me. So I took his advice and went to my Master; and he kept me up to his home house, from Saturday until Monday. then he took me down to the plantation again. Now the overseer pretended that he was very glad to see me, but I kept my eye on him. So the overseer went to the house with the Boss man, then he come back to attend to my case.

Now I was an expert wrestler. So I had made up my mind to fight him. now just about while I was thinking about all of this I casted my eye across the field and I saw the over seer coming with his cowhide under his arm. so he came direct to me and said where have you been young man. I said I've been in the woods. I spoke very saucy too. he looked at me, and said pull your pant down, and come lay down across this cotton row. I will show you how it is to run away. I would not obey him, he said don't you hear me sir. I didn't obey him yet; so he caught ahold of me. of course I was afraid to fight him openly. so I pretended that I was trying to get away from him and fighting him on the sly. Now as I said I was an expert wrestler so I could throw him as fast as I please. He was fighting all he could at last I het [held?] him so. He caught my ear in his mouth, and began to bite me.[6] so I told him that he was not to bite, I had forgot he was the overseer.

Now we fought about two hours,[7] and he could not do any thing with me. finally I got away from him but my pants fell down around my legs and I fell, and when I did he fell upon me so we went at it again. Now there was about thirty people in the field including women, all picking out cotton and they was afraid to hold up their heads for fear of the overseer whiping them, so I saw the overseer kept looking over towards the men; so at last he called for jef, one of the men to come to help him to tie me. he said I didn't mean to whip this little rascal much, but he would not obey me. I have been sick or I could tie him my self. so jef a very strong man could tie me himself without the aid of the overseer. so Jef held my hands and the overseer tied me, and he hit me ninety five lashes in all, he would give me more but he was afraid that the boss man would hear of it, so he stoped at that. Then he let me go. Now during the evening, he went over to visit his neighbor overseer and he said, why Horton, for that was the overseer's name, how come your face all scratched up. Horton said is my face scratched up? The other one replyed yes. Horton said why that little Wallace run away and I went to whip him and you think he didn't fight me? The other over seer replyed he was a marking you. This ends my first run away.

Now Mr. Horton had a very nice looking woman which belonged to him. The overseer all ways had a head man over each Squad of hands. Now the following Spring about a dozen of us was ploughing in young cotton.

There was an old man by the name of Sam and he was in love with Sarah and wanted her to plough with him and

Madison the Headman wanted her to plough with him. So Sam
beat him and it made him angry. so he found fault of our
ploughing for revenge and said he would tell the overseer that
we covered up cotton naming so many that had covered up
cotton bringing my name in also. The overseer was gone to
town that day and returned that night. when he came Madison
told his story and the overseer said that he would whip everyone
that covered up cotton. now when he said this I was standing
close by him and when he saw me he commence to talk about
something else but I had heard what he said. then I helped to
finish feeding the mules and went to the house and got me a
drink of cool water out of the old well. Then I got over the
fence to see what would be the result.

So while I was over the fence not very far from the
overseer's house, I heard him say, Now you go and fetch Wallace
to me and I will whip him; and then I can get the rest of them.
Now these people was very eager in looking for me but I
escaped them all and went on my way and slept in the woods
that night on an old log. So I stayed around the plantation a few
days and then I took my departure. I did not go to my Master
anymore for protection but I tried to find refuge in a strange
country which I did.

Now Columbus Miss was a about Twenty seven miles from
the plantation I lived on. This City lay right in the way I went
and as I was on my way to Columbus, it being the month of
June and the leaves on the trees was very thick. and as I went on
the road I came to a place which was in an elbow form, so much
so that I could not see anyone until they got very near up on
me. Now right at this place I met an old white man with one of

these long old shot guns and he says who do you belong to boy.
I said to him that I belonged down in Pickensville. I did not tell
him who I belonged to. he told me stop. I told him that I did
not have time. he said if you don't stop I will shoot you and
from that he snatched his gun from his shoulder. I jumped in
the thick bushes but he fired in an instant and the shots went by
my ears and shoulders cutting the leaves like so many hornets
but none of them hit me but they made a bad noise. So I got
by safe. then I went until I got to Columbus but I did not stay
around there long. I had to keep concealed all of the while. this
was the furthest place I had been. So I thought I would go back
to Pickensville again. so I did but with much more dificulty than
I come with. Now in those days I being quite wild and half
starved part of the time, I was a very fast runner. I did not mind
a man except he had a horse or a gun. So I got to walking in the
public road on the week days as well as Sundays. But I soon was
caught up with.

Now when I was on my way from Columbus to Pickensville
I thought that the road being easier than the woods, I would
walk in the road. So I got along very well until I got about Ten
miles from Pickensville. I come to a house that was right on the
road no more than about five steps from the road. and upon the
porch sat one who owned the place by the name of Joseph
Osbern. he was a very tall man and a very fast runer. so he saw
me pass and said who do you belong to boy. I did not tell him
who I belong to but I told him that I belonged down in
Pickensville. he said stop and let me see, supposing that I had a
pass. so I would not stop and he said if I did not stop that he
would set the dogs on me. then he began to clap his hands and

call the dogs but the dogs was in the field where his people was aworking. now as I set out to run I thought first to run down the road.

Now on either side of the road was a field and the road passed through this way and the road had a fence on each side. Now the woods was a good way off and before me in the way on the road on which I intended to run down, there was a man coming on a horse and also a man with a stage which carried the mail in that country. Now I saw that I was near being caught. And the man was a very fast runner. I throw myself out in my best running. I jumped over the fence and run so fast through the corn that I could not see. Now the roasting ears would all most knock me down I was running so fast. Now while I was running I heard this man say catch that runaway, then I looked back and I saw that I had left him about a hundred yards behind me.

Just about this time I run in the midst of his hands working in the field. Now the dogs was there. Three Dogs took right after me and the people throw their hoes at me. I had on a very heavy coat and I could hardly keep it on. I was going so fast, the dogs nor the people could catch me. so as I was runing I come to a very deep ditch. I did not think that I could jump over the ditch but any how I tryed and the Lord helped me and I jumped over safe. and the dog that was nearest to me tried to jump over too but the dog fell in the ditch and could not get out so I lost them all. if I had fell into this ditch I would been caught, it was a very wide deep ditch. Thus I got by them all safe.

Then I went on until I came to Pickensville. Pickensville is a village about Twenty three miles from Columbus. Many of those rich farmers lived there and kept stores there, a very nice village. When I got there I went to a friend, which attended to my Master's church and asked him for something to eat and some place to stay. so he give me some thing to eat and told me that I could stay up in his church steple and he would give me something to eat as often as he could. I accepted the offer very gladly for it was much better than laying about in the woods. Now I stayed up in this steple some time and I suffered greatly from the hot weather as the sun beamed upon the steple in those hot days in the summer in Alabama.

Now I had a quart bottle to keep my water in to drink but it being so very hot it seemed to me that I would famish before night come to get more water. I could see my Master when he passed by every day going to his store but he did not know that I was there. Now I used to come down every night to receive what my friend had for me to eat so one night as I little thought that anyone would come that way and it being a late hour, I had not been down more than five minutes and had just got over the fence when I heard some one coming. so I layed down.

Now I had an old straw hat. it was a white one, and that old white straw hat betrayed me. so the man come along and looked over the fence and said who is that, but I thought that he could not see me as I was laying so close to the ground but the man kept his eye on my hat and jumped over the fence with a Pistol in his hand. then I tryed to run but the man was to close on me and told me if I did not stop he would shoot me. so I could not do any thing. I was caught. I had to give up and go with him. he took me to my master that same night and they kept me secure that night until the next morning. then my Master said that he would not send me to the plantation again, but he would keep me up to his home house to wait on his private table, to work in the garden, to black his boots, and to wait on him in general. but they never found out that I stayed up in the church steple. So this was my second run away.

Now Mr. Chalermers was a very hard man to get along with. he use to whip the people about his home house very often and severe; and it was thought among the people down in that part of the country that if a servant would run away from the over-seer, he would not run away from his master. but I soon showed Mr. Chalermers that I would run away from him just as soon as I would run away from the overseer. Now I attended to his store and fed the cow and horses and sawed wood and made fires and went errands to different parts.

Mr. Chalermers wife was also very hard to get along with. she had me whiped several times. Now they thought that they had broke me up from running away. so one day she had a nice whiping laid up for me and said that she was going to make my master whip me When he come home at night. so I got tired of being whiped so much. so I took what little things I wanted and took my departure.

At this time it was beginning to get cold. it was the latter part of November. I made my escape to the woods all right and safe. Now I went about nine miles up the country but I suffered very much from hunger. I went as long as four days without anything to eat but one hickery nut that the squirrels did not get.

Then I went a little farther up the country and got acquainted with a man and made an agreement to teach him for something to eat. so I teached the best I could for some time and this man would come out once or twice every day to get his

lessons. But after a while some of the other colored people saw
me there and they made so much noise about it that my friend
told me that I would have to leave, that I could not stay about
there any longer for he was afraid that I would be caught. So I
took my departure one cold Sunday night when it was high
water time and it was difficult to get over every little branch.
However I went all of that night towards Columbus and the next
day I got about two or three miles from the city where I was
detained on account of the guards that guarded the Bridge over
the Luxapatilla river. so I layed around there in the woods Three
or four days and I made many attemps but I was unsuccesful.

However one day I thought I would try another plan. so I
went down to the river side and coming up the bank of the river
easy I succeeded in getting up to the guard house where I could
see through the cracks of the house and see what they was
eating and could hear what they said they would have for
breakfast the next morning. By this time a cow walked upon the
bridge and one of the guards said to another, John see who that
is on the bridge. so he went to the door and looked through the
bridge for he could look through from one end to the other.
Now all of this time I was standing behind the house hearing
all they said. so by this I know that they were all in the guard
house. so then I sliped off my shoes and went across the bridge
as easy as I could. the bridge was covered over with a house.
Now frounting [fronting] the bridge on the other side was a
breastwork thrown up. so when I had got over the bridge and
was going around the breastwork the guard come on his post
but I got by all safe.

Then I went on until I got up in the city. I got in a little
after eight Oclock. Now after nine Oclock at night the police of

Columbus would take up any colored person they seen walking the streets and put them in the guard house if they had not a pass to show their liberty. So I had been to Columbus two or three times before and I know a family there. so I went to their house and knocked at the door and the old Lady told me to come in. so I went in and told the old Lady that I run away from Pickensville Alabama, Naming the people I had run away from and asked her could she give me anything to eat. she said yes. so she give me a good supper and said Son, you must try to get out of the city before nine Oclock or you will be took up. Now while I was offering my thanks to the old Lady, the clock struck half past eight. Then she wished me God's speed and I bid her good bye and went out to find my way out of the city before nine Oclock. so when the clock struck nine I had just got out of the city and struck the woods near about the road that led to Aberdeen and as I had never been through the city that way before I did not know whether that was the right road or not.

So just about the time I was thinking what to do I heard someone tearing the bushes fearful before me. so I hid behind a large tree to see who it was. the moon was a shining dimly so that I could see a little distance before me. Now the place where I was had a little slant which led from the road and there was a gully where the water run down when it rained. so I saw a man coming down this gully and I hid my self good behind this large tree. so the man came down the gully until he got opposite to me though I was on the opposite side of the tree to him. he said good evening, so I said good evening. how he saw me I don't know however I wanted to see some one very bad. Then I said to him is this the right road to Aberdeen? for the road was in sight, and he said it was but said he how are you going to get

there. there is guards on the road said he. Then I ask him if
there was no way to get around them, he said yes. so he told me
how every guard stood up that road for nine or ten miles or as
far as there was any and how to get around them all. who he
was I do not know nor did he ask me who I was nor where I
was going to nor where I come from.

So I went on until I got to the first guards. long before I was
near to them I could see their lights, then I went around them
as I was directed. so I got by them safe. Now about six miles
from the first guard was an old steam mill and there was guards
there also. Now this man told me when I got there I must go
about a mile to the right of this mill and go around it and come
in the road bout a mile below it. so when I got there it was very
late in the night and on the side that I was directed to go was a
thick woods and on the other side was a lot of stables. so I
thought I could pass on the side that the stables was on but
the first thing I know I was on a soldier's beat. but I had the
presence of mind to lay down and crawled on my all fours and
made my escape. Then I went the way that I was directed and
got by safe and this was the last place of guards that I was
informed of.

So I went on with all ease for about a half an hour. When I
was passing a large plantation a man run up to me and said are
you the one that stole my geese last night. I told him no and
told him where I had come from that night. he was a colored
man and I suppose he was the headman of the place. someone
had stole his geese the night before and he was out on the
lookout for them. So I got by there safe.

Then I went on until I come to a hill where the road
ascended and it soon des cended. you could not see any thing on

the opposit side of the hill until you got on the descent of the hill. the road when going down the hill turned in a croocked way. so when I got near about the foot of the hill on the other side, I saw about a dozen tents, as a camp of soldiers, and I know that the man I saw did not tell me anything a bout that place. I was struck silent at the sudden appearance of the camp and moreover I saw Two men come walking direct to me. I could not run for they was to close to me. I was badly frighten when they came up to me. I found that they was colored men. then I asked them was there any white man there. they said yes, but he is asleep said they. then I asked them how the country stood there but they did not know anything about the place at all. So I got by them safe and went on. I could not travel much the next day, not until night. the next night I went to a house and knocked for something to eat and there came a woman out to see what I wanted. so I told her that I wanted something to eat but she declared that I was her master's son, trying her to see whether she would give a runaway anything to eat or not. I told her that I was not but she would not believe me. she told her husband and I had to get away as fast as I could. so I could not get anything to eat there.

So I went and slept in an old gin house[8] that night still hungery and could not get anything to eat. the next day I seen a woman cuting wood off of a stump and I called to her to see if I could get anything to eat. she said if she was home she could give me something to eat. she said she was there siting up with a sick man. I asked her where did she live. she told me. so I went to her house or to her owners house and went in the yard and found a large dog there. I drove the dog out of the yard. he did not bark. Then I went in the kitchen and found a nice looking

dish safe. I tried to get in it but it was locked. then I took hold
of the bottom of the door and fell back on it and it broke half
in two. then I run out to see if anyone was coming but I found
the dog upon the fence looking to See what I was doing. so
I run him off again. then I went back in the kitchen to see
what I could find to eat and I run my hand down between the
broken door and drawer and I found a large dish of chitlens
[chitterlings]⁹ and a lot of sugar. Now I was very near starved I
eat near about half of them then. I looked for some bread but
I could not find any but I found some meal. so I cooked me two
sorts of bread, one plain and one sweet.

Then I looked around and I found a wallet and I put sugar in
one end and bread in the other. Now the day that I saw the
woman I layed round the ferry that crossed the Tombigby river
and there was some wagons crossing the ferry who was moving a
man by the name of Gardener. so I heard them tell the ferry man
who they was a moving. Now when I got my wallet full and had
eaten all that I wanted myself, it was near about day light so I
had to get from that side of the river before the people come the
next morning. I went down to the river, it was high water time,
and those large steamboats could run as high up as Aberdeen.

I went along the river shore and I saw a boat pulled up on
the shore and tied to a root and I saw the tracks of two men
which I suppose was hunters. I untied the boat and rowed
myself across the river but the water was so swift that I could
not land where I wanted to but I landed at an old steam boat
landing place. and I got out and pulled the boat up on the shore
and went my way but I soon got into trouble.

Now there was many little branches leading to the river. and
they was very difficult to get over. I had nothing to walk on but

a small log from which if I had fell I would never been able to
get out of the water. it was deep and the banks was steep. if I
had fell in I would have been drowned. I had no more than got
over this before I saw a white man before me and he saw me. so
I had a large stick in my hand and I did not intend that he
should take me. when I was a good way off I said good morning
sir; he said good morning, then I asked him, had he seen any
stray hogs he said no, who. do they belong to. I told him that
they be-long to Mr. Gardener that was a mooring from that
side of the river. so he know that that man was moring from
that side.

I had my wallet in my hand as though I was hunting for
hogs. he asked me what kind of hogs were they. I told him that
some was white some black and some spotted. well he said, go
back and see if any of them are they, pointing at some at a
distance. so I went back and was glad of the opportunity. Now
on that side of the river was all cleared land and there was no
way to escape except going along the river. at this part of the
river the banks was very steep. the bank I think was near a
hundred feet high. I had to make my escape between the
channel and this high bank. all that I had to hold onto was little
rook and bushes that grow in the bank. I had to go very fast and
it seemed to me that every particle of ground that I took my
foot off of on this bank rolled down in the channel of the river.
I heard the man hollowing after me but I kept on and got by
all safe.

Then I went and layed in sight of a large plantation the
balance part of the day and that night I went up to this
plantation. I agreed to give them sugar for meat which they
very gladly accepted and I give them sugar and they give me

meat. of course they did not have much to give. I slept in an old barn that night and the next day I slept very late and when I awaked I saw three or four good size white boys with Clubs in their hands. I suppose they had seen me up there and was planning out how to take me. so I got down and run off and they did not see me. so I got by all safe.

I was then on my way from Columbus to Aberdeen and that same night I got to a plantation which was about five miles from Aberdeen. So I went in and knocked at the door of one of the houses and asked for something to eat and the house that I knocked at had about four or five men in it, all colored men. they told me to come in but I was afraid to come in. but they was all friends to me though they had never seen me before they told me that they would not let anyone hurt me while I was in that house not even the overseer. I told them that I was a runaway though I was only a boy. they gloried in my spunk.

There came someone several times and knocked at the door to see who I was but the men told them to get away from there. they give me all that I wanted to eat and saw me away from that place all safe. so I went on until I got to Aberdeen. I suppose that I got there about nine Oclock that night and I came to a door that opened in the street. I thought probably that it was a carriage house. so I knocked and then I looked through the cracks of the door and I saw that they was officers. so I got away as soon as I could.

So I on trying to get up in the city, I climbed over walls and went through yards. At length I got to a yard which my mind led me to go into. so I went in and saw a woman in the yard taking in clothes, and when she saw me she ran and fell in the door supposing that I was a spirit. for they had a very severe dog and they thought that no one could get in the yard without the dog finding them out. Now when the young woman run so excitedly in the house, the old lady, her mother run out to see who I was and she said for God's sake child come in the house. it is a wonder said she that the dog has not torned you to pieces. so I went in and told the old lady my circumstances and she said I have a son that has been away for some time and I expect him home soon and I expect some visit-ors here tonight. and when they comes I will tell them that you are my son. so she give me something to eat. So bye and bye some one ranged the door bell. these people that I am speaking of was servants[10] and those that come to see them was servants so they had them in the house. So the old lady introduced me to her friends as being her Son. They was very glad to see me. Now see what a responsibility she took on her self for she had never seen me before. She kept me there all night and give me a place to sleep.

Now the dog made a great noise when I was in the house as though he was bad. he kept it up some time. The next morning the old lady give me something more to eat, and hoped that God would help me on my way. Then I went out the way that I come in and went around the rebel camps and went around the city and went on my way. I was now trying to find my way from Aberdeen to Okolona. so I went on until I got to the prairies which gave me a great deal of trouble to get through. and the day that I left Aberdeen it was sunday so I went on until I come near the Mobile and Ohio railroad.[11] and while I was laying down by the side of the fence suning myself, there was a colored man and a white man running a rabbit and the rabbit run right along by me and they seen me. And the colored man said to me are you the one this gentleman hired the other day, speaking of

the man whose place I was on, I forgot his name. I told him yes.
he asked me how did I like my new home. I told him very well.
so I got rid of them.

That same night I met up with an old colored gentleman
and asked him for something to eat. he took me in his house
and give me some thing to Eat and kept me in his house all
night. The next day I went and lay in an old church all day in
sight of Egypt station. And that night I went to another
plantation three miles from Okolona and went in there and I
found another friend and he give me something to eat and told
me how to get around Okolona. so when I left this man's house
it was about two Oclock in the night.

So I went on until I came to the city. Now the man told me
that I must go the right of Okolona and I would get by all right;
and when I got there and had almost got by, I came to a house
which had no wall around it and I was getting quite bad off for
some clothing. So I was looking around among the barrels to
see what I could find to wear. so while I was there I heard
something coming, runing like so many hogs runing when they
are frighten. I always carried a large hickery stick so I jumped
back with my stick to defend myself and behold there was three
large dogs fearful to look at. They did not did not make much
noise but each dog tryed for himself to take me. all three of
them stood at respective distance. Now I had to work very hard
to keep them dogs from taking me. I had to fight all around me
to keep them off. if the Lord had not helped me I could never
outdone those dogs.

Now while these dogs was a fighting me, the man had time
to get up and dress himself. but when the dogs could not take
me themselves, they could not be set on. the man asked who

was there but the dogs did not renew their attack. so I got by safe. Then I went on up the railroad trying to get to the free states. I thought being I had got that far I could go further. so I went on until I got to Shanel Station Eight miles from Okolona. I did not get up there until night about eigh[t] oclock, and I was geting very hungry. so I went up to the back of an old kitchen and looked through the cracks and I saw a colored woman in there and asked her for something to eat. so she pretended that she was afraid of me and run in the house and told the white people.

By this time the dog had found me out. he had me bayed up in the chimney corner out side. by this time the man came out and caught me and took me in the house; he also asked me who did I belong to. I told him. he asked me where I was going to. I told him that I was trying to go back where I came from. he asked me where did I come from. I told him that I came from North carolina. So he got the News papers to see if he could find any such a boy as me advertised in the papers. so he could not find any to correspond with me. then he asked me who came with me. I told him, no one. so he did not believe that I came from Pickensville Alabama and said he believed that I belonged to another man that lived down in the prairies. I told him that I did not. so he said I will make you tell in the morning. So he kept me secure that night and the next morning they took me in an old store and asked me the same questions but I told them I did not. so they said that they would whip me until I did say that I belonged to the man that they said I did. so they began to whip and they het me so I soon told them I belonged to the man but I did not. and I thought also that I might make my escape from them before they got ready to take me down.

So they stoped whiping me and the man that caught me took me back to his house again and it rained all day long. so in the evening the man went away and he thought of tieing me but I told him that I would stay and when he was gone his son went out to cut some wood and his wife sent me out to bring it in. the yard was all paled in and the paling[12] was very high. so his son went to feed the horses and hogs &c and when I had brought the last in, I went out as though I was going to get more and jumped over the paleings and made my escape to the woods.

Now as I said that it rained all day, it seemed that all of the little branches had become so deep that I could not get over them. so I stoped under an old house that night.

Now the day that I was whiped in Shanel[13], there was two colored men there who belonged to a man that was the boss of the railroad and they lived up the railroad not far from shanel and their house was close by their master's house. of course I did not know that there was anything up. So when I came out from under this old house I was quite cold and stiff it being in the month of January. so I goes to this house. being the smalles[t] one there, I thought that I would find colored people in it and so I did, but they did not befriend me. They recognize me as soon as they seen me and said why did they let you go. I told them no I got away, thinking they would be like other colored people that would try to help me, but they rose up and caught hold of me and said you can't go any further. so I tried to fight but the first thing I know their master jumped in the door with a pistol in one hand and a bowie knife in the other. he had the pistol cocked and his finger on the trigger. he put it in my breast and made many threats to shoot.

I had on very thick clothing on my breast. he stabed me two or three times but the knife did not go through far enough to hurt; so he knocked and kicked me all the way to his house and when I got in there I seen another runaway a laying on the floor which they had caught the same night all wraped up in chains and locked around the neck. so this man beat me severely until daylight, sometimes with his cowhide and sometimes with his gun. so he thought he would make a final end of me. he took me by the shoulders and back of the neck and beat my head on the bricks under the mantel piece until the blood run down on the hearth; and when he had beaten near about all of the sense out of me, he then threw me [into the] back part of the fire. I got my hands burnt badly, but I jumped out as quick as I could. then he kicked me across the fire and asked me if I had the assurance to jump out of the fire when he put me in there. so he sent his boy to bring him another chain and a lock and he throwed me down in the floor and locked me with the other man. then he told me that I should be hung by Sun up.

Now it began to snow and got quite cold for that was North Missisipi. So they had their hand car there to take me down to Shanel to be hung. So the men went off to talk the matter over and I was reprieved and sent back to the house again and now to make sure of me they put hand cuffs on my hands. they put hand-cuffs also on the other two men for I was the third runaway they had. So they hand cuffed us all three together and then chained us together. And when they fed us, they would cut up our food for us because our hands was hand cuffed.

So this man kept us four or five days and on Friday he took the hand cuffs off of the other two men's hands to take us home to our ow[n]ers. These other two men lived in Missisipi the state that they was caught in but I lived in Alabama. Now when he tried to get the hand cuffs off of my hands they was so tight that they could not get them off. they took me down to Shanel to get them off but they could not. but the man had so repented of what he had done to me he said that he did not want to take me along with hand cuffs on, and the other runaways at liberty so he went back and waited until the next day which was Saturday. So he tried to get the hand cuffs off again and when he put the key in the hand cuffs unlocked without any trouble.

Then we started for Okolona with the hand car and arrived there in the evening. but there was no train up until Monday. the Yankees had burnt all of the bridges beyon[d] Okolona. So the man had to keep near all night Saturday night and Sunday night to watch us for fear that we would get away. So everything passed off very well. the train come up Monday and we went down to West Point and stayed all night; the next day we took the cars to Artesia and from there to Columbus and stayed there all night. we stoped at the city hotel. the next morning we took the stage to Pickensville Ala.

Now that was where my Master lived. he took me to my Master's store and he was very glad to see me. he asked the gentleman where he caught me. he told him. So we went to his house to get our dinner and he asked the man how much did he want for bringing me home. he then asked my Master could he send him back to Columbus again that night. he told him yes. then he told him that he would only charge fifty dollars for me. So everything passed off very well that night.

The next morning he called me into the house and asked
me to tell him something about my runing away. I told him that
I was trying to get back where I come from. so he told me that I
could never get back. so he took me and give me about Twenty
five lashes and sent me down to the plantation. he did not keep
me at his home house any more and This overseer that he sent
me under was the second overseer that I worked under on that
place. This ends my Third runaway.

Now the Second overseer's name was Scoggen. he was noted
for his smartness for he said that a slave could never run away
from him with out being whiped; he meant by this that he
would never threaten them before whiping them. if he had
anything against us, he would come and take us by surprise and
whip us. So he said that he could break me up from runing
away. They thought that when they told me that they was going
to whip me that it always frightened me and made me run off.
So they thought if they whiped me without telling me before
hand that it would be all right and I would not go.

So there was the grounds that Mr. Scoggen thought that he
would break me up from running away on. Now I lived with
Mr. Scoggen or rather worked under him from the latter part of
February 1862 until Harvest the same year. Mr. Scoggen whiped
me once during that time with his walking stick, he hurt me
very much. he came on me and whiped me without telling me.
so I took it but it went very hard with me. I could not do any
thing but run away. So I said in my mind that I would see when
he whiped me again.

So a few Sundays after that he sent me and another boy
after the Mules and Oxens. the boy's name was Andrew. So
Andrew brought the Oxens and I brought the Twelve mules and

Andrew he had Six oxens to bring. you may know that these oxens and mules was very rulable [valuable? Controlable?] that we two boys could bring each drove separate. So I got the mules home all right and Andrew in bringing his oxens along the road that led through the woods there got a wild Bull in among them and went in the lot with them. so the next morning the ox driver came and got his oxens and left this bull in the lot. it being dark he did not know that it was a wild bull and this bull stayed in the lot until breakfast time. and when we all came to breakfast the overseer sent one of the boys in the lot to drive the bull out but the bull was so enraged that he took after the boy and like to have caught him before he could get over the fence. then they opened the lot gate so he would go out but the bull was very contrary in going out at first. so at length he darted out the gate and took after one of the men and run him about fifty yards. the man was a fast runner. the bull was about a foot from the man with his head down looking bad enough but the man run towards the gate that led to the road and jumped up on the fence by the gate post and the bull run out in the road and went in the woods. the gate was ready open for the bull to go out.

Now at this excitement the overseer become very much ouraged [outraged] and said where is that Andie and Wallace. I am going to give them a genteel whiping this morning. so I said you are not going to whip me this morning, but I tell you that I had to do some tall runing after I said that or else I would been whiped. Mr. Scoggen had forgot that I was such a soon runaway. However I beat his time and got away safe.

So I went up to Pickensville and seen the wagoner and made an agreement with him to take me across the bridge at the Luxapatilla river. this bridge was guarded and I could not get across so easy as I did before. So I left for Columbus the over night, and the wagoner left the next morning. he over took me the next evening about five or six miles from Columbus and I got in the wagon and got over the bridge by the guards all right.

Now this wagoner was the main wagoner of the place. he hauled goods to different parts of the country, round about for forty, fifty, Thirty and Twenty miles. so when he got across the bridge he camped for the night with the other wagoners for they could not go up to the city until the next morning if they did not get there before night.

So it was costomary for wagoners to have wagon boys in those days. So one of these other wagoners wanted to sleep in the wagoner's wagon that I was in and he found me in there and asked who I was and he told him that I was his wagon boy and that I was sick. The wagoner understood my move. Now between midnight and day, I got out of the wagon easy and made my escape and no one heard me. So I went around Columbus and went on toward Aberdeen. Now I had to get to Okolona before I could take a nights rest, it was about a hundred miles from where I started from. So I went on without any difficulty, only I was very sick from eating so much fruit, until I wanted to cross the Tombigly river. the water was not very high only I could not get across without swiming. I waited all day for an opportunity. at length there was a colored man came rowing a boat down the river. so I hailed him and asked him to take me across the river and he seemed to understand my condition without my telling him anything at all; he told me to come lower down the river where no one could see me and rowed me across the river and asked me no questions at all. I thanked him and went on my way.

So I got to Aberdeen all right and met no opposition
whatever and got by all safe. Then I took my course for
Okolona. I went on all right and met no danger though there
was many dangers in the way but I got by them all and I got in
about Three miles from Okolona and saw my friend again and
got his directions of the way and got something to eat and
stayed until morning. So I got to Okolona just about sun up
and around the city about a half an hour after the sun was up
and now I had traveled from Wednesday night until Tuesday
morning without taking one nights rest. So I was completely
exhausted.

So I layed down in the cornor of the fence, in sight of the
rebels. They was passing all of the while. Now I slept from
Tuesday morning until Wednesday night. Then I felt like I
could go on my way. So I went up and saw a man and asked
him for something to eat and he give me some bread for he had
no meat to give. So I asked the man for the right road that leads
to Corinth and he showed me the right road. So I went on with
all ease traveling the road some times and some times on the
rail-road.

Now I was from Wednesday night until Sunday morning,
getting up in about two miles from Rienzi, not far from
Corinth.[14] Now the Yankees had been through this part of the
country once, and it was very hard for me to see any one in
these parts that I could trust. There was no colored people
there. so I got quite careless; being I couldn't see any one, I
thought that I was out of danger. so I walked boldly upon the
railroad every day; and on Sunday morning as I went upon the
railroad, it happened that a man rode upon the railroad the
same time that I walked upon it. no doubt he was a rebel scout,

the road being straight they could see a long distance each way. So as he rode upon the railroad on his horse he seen me just about the time that I seen him. So when he seen me he turned his horse into the woods in an instant, so I did not think anything of that. I went on but not upon the railroad.

So by and by I heard some dogs barking behind me but they barked so irregerlar that I thought that they was some ones yard dogs. Now there was stationed in that country a man with hounds to prevent runaways from making their escape through the lines to the yankees. but I did not know that they was there. and this man that seen me on the railroad had been and informed them of me and they had no trouble to get on the track of me. So the first thing I know they were in sight of me and the man that was with them seen me. So I run about two miles right through the woods and of course the man could hear which way the hounds was going and they know the country and rode around on their horses and cut me off.

Now while I was a runing, I run out of the woods of which I was in all of a sudden. and when I did, I saw one of these men about five steps from me with a cavalry pistol in his hand with dead aim on me and asked me wasn't I going to stop. I told him I was but I kept moving a long. he still kept aim on me and said you don't look like you are going to stop; he said I thought it was, a man that I was after, I didn't know that it was a boy. he moreover said I don't care whether I take you or not. I will make my dogs eat you up; he said that because I kept moving along. then he got off of his horse and tied him to the fence.

I could see some more rebels off about fi[f]ty yards to help see the end of me if I did not stop. So he got over the fence and untied his bull whip from around his neck. The same time the hounds was stationed all around me waiting for the word. so the man walked up to me and struck me one lick with his bull whip. then he told the dogs to take hold of him. So every hound jumped on me at the word and they carried me about like a feather. there was seven hounds and they was large and all of them had a bite at me or in plainer words they all had hold of me.

Now these dogs bit me all over while some of them had hold of my legs the others was biting me on my arms hands and throat. I could not fall for while some of them would pull me one way, the rest would pull me other way. he made them bite me four or five minutes. Now these hounds was so trained to bite without tearing the flesh. any one might think after biting four or five minutes and seven of them at that, that they would have torned the flesh to pieces, but they would not tear the flesh out. So when he thought that they had biten me enough, he made them stop and then he asked me who did I belong to. I told him.

So he took me to his house and the other men went their way. he wrote down to Pickensville Alabama to my master that he had caught his boy Wallace and if he wanted me, to come after me or send for me. but my master was afraid to come up in that country on account of the Yankees. So I stayed with this man and worked in his field. he did not confine me. he know that I could not get away and I was only a boy. he was not afraid of me doing any harm. So about three weeks after I was caught the Yankees made a rade [raid] through the country and went way below me or the place that I was at. but two rebels had me an[d] another runaway which they had caught in the like manner as they caught me, guarded in a thick woods. The other runaway was a large stout man and they had iron Bands on his legs. So the news of the yankees coming down the country excited him so that he would not come up to see me, not until he heard from the man again. so he wrote to the man that I belong to and told him that the yankees was all gone back up the country and if he wanted me he must come after me.

The Yankees had all gone back up the country. And so I was kept at this man's house from August until October [1862] before he came after me. so at length my Master came after me on Sunday evening and stayed all night; and that same night the yankees started down the country again, but however, he took me away the next morning. He came after me on an old mule and so my master rode part of the time and I rode part of the time, the mule was so stubborn that we could not get him out of a walk. We got as far as Borona that day and stayed all night. they had me chained to my master's bedstead that night for fear that I would get away again, and the next morning we took the hand car for Okolona. we had fifteen miles to go. we got there by half past ten Oclock, in time to connect with the train for Mobile. So we went down to Artesia and there we connected with the train for Columbus. we arrived there about four Oclock in the evening. So he hired a road wagon and a driver to take us to Pickensville Ala.[15]

Now as my Master was taking me home, we had to pass by the rebel pickets; and they wanted to know where he got me from. So he told them that I had run away from him and I was caught up the country and he had been up to bring me back. They wanted to take me and make a Target out of me. But my master said no, I belonged to him but they said Mister what good is this boy to you if he is all the time runaway. So while one was talking to my master about it, the other one was carrying me to tie me up to a tree for a Target, but my master told them that he could sell me so they let me go.[16]

So we arrived at Pickensville that night. And everything passed off very well, he did not say a cross word to me atall. the next day he put me out in the garden to work. in the far south you can work quite late in the Fall in the garden. So I worked in the garden from Tuesday morning until Friday morning. So he told his man to hitch up his carriage and he came by the garden where I was aworking at, and told me to get ready to go with him, that he was going to take me away.

So in a few minutes after we started and went over in Mississipi to the Mobile and Ohio railroad to a station by the name of Broocklyn.[17] And we took the cars that evening and arrived in Mobile the next morning. so he took me to a trader's yard but my name was so well known for being such a runaway that the trader would not let me stay in his yard. he came to the yard and said boy come out of there, I wouldn't buy you. I said why sir. he said I hear that you can go to the Yankees anywhere. I told him that I was not trying to go to the Yankees. I was trying to go back where I came from. he said, where did you come from. I told him. Then another one of them said why he is an Indian. but the trader said no he is a Portugee.[18] So he took me to another trader and this trader put me to work in his store. they would not let me be put in jail for fear that I would tell some of the rest of the colored people something about the country. So my Master returned home again and left me to be sold. So this ends my fourth runaway.

So I was kept in Mobile about two months at one of the clerks house, to wait on his family before any one would buy me. At last there was an old Gentleman who wanted a boy to wait on his family and to drive and to do general house work. So the old gentleman heard that I was for private sale. he made an application for me and bought me. The old Gentleman's name was —Mr. Minge.[19]

So the clerk came home to his dinner and told me that I was sold; the next day I was sent down to his office and in the evening he took me out to his house. I waited on his table, drove his Carriage, ploughed in his field and Garden and hauled wood and went Errands. Now the old gentleman liked me very much because I was such a handy boy to plough and work in the garden. I had worked in the garden so much that I know all about it. But his wife and Daughters wanted me in the house. So just as soon as the old Gentleman would go to his office in the morning, they would call me out of the garden to work in the house. Anyhow to make my story short, Mr. Minge had an old pair of harness which broke very often. And harness at that time being very exspencive he wouldn't buy any. So the harness broke several time and was mended again by me and him, until they became nothing but Patches. The Horse too was very touchy, he would jump if you touch him with the whip.

So I got along very well with the old gentleman until the latter part of August but I did not get along with the rest of the family. Now as I said the harness had broke several times, they now broke so often that the man threaten me. So he said young man if these harness break again I will whip you. so I told him that I could not help it and he said I hit the horse and made him do it. but the horse was very spiteful; he run away with Mr. Minge once, and he know how quick he was.

Now I had already joined the Baptist Church in Mobile and thought as I had had no cause to run away I would be baptized. but in those days the Rebels would interfere so much with the colored people when they went out to baptize and make them work on fortifications and boats all day sunday that they postponed it. So long about the latter part of August there was about Twelve Thousand of Rebel Soldiers in Mobile being there was not much fighting going on. Now this was the August of 1864. Mobile was in quite a stir. it was most imposible to get along there was so many inhabitants without the soldiers.

After I had been with Mr. Minge nearly a year[20] he was quite harsh to me at times. The Old Gentleman give Twenty Hundred Dollars for me. I did not serve him quite two years. So one day I was driving him down town as usual. And he was telling me again what he would do to me if those harness broke again but he did not know that I was a runaway and that I was sold for runing away.

Now as I was on my way back up town as I was crossing the car track in Dauphin Street, I hit the horse to get out of the way of the car as quick as I could. the horse jumped all of a sudden and broke one of the traces and the shaft on the side the trace was on fell down and hit the horse in his fore flank. So the horse ran away and I could not stop him nor get out. So the other shaft got out of the strap and came down and the ends of it hit against the rocks in the street and that broke the other trace a loose and out he went like a locomotive. I had to let the rein go or else he would pulled me out with him. Now he was going in such a speed, when he left the carriage it kept on at a rapid rate neither could I get out yet. Now the wheel struck the car track and the carriage was going so fast, the force of it sent right across the track in front of a large store which was open. the shafts struck the curb stone and the force of the carriage wound them all up in splinters and dashed me against the spatter Board of the carriage. And the Rebel soldiers said to me, boy you are blessed.

So they soon caught the horse and brought him back and put me upon him and they pulled my carriage to one side. And I went home. So when the old Lady seen me coming on the horse without the carriage she was anxious to find out what was the matter. So when she found out that the horse had run away and had broken the carriage, she got very angry and said it was my own carelessness. She commence to say what Mr. Minge would do to me. So I got angry and spoke very short to her and told her that the harness was no good and that I could not help it and they could do what they please. Now this was counted the height of impertinence of a slave in the South. Now she thought that I might run off. She put me in the cellar to lay Bricks that is to lay a brick floor and she stood in the door and sent the old colored man down town to get my master.

Now while he was going her next door neighbor was taken dangerously ill and sent a servant to tell her if she wanted to see her alive come at once. So this message took such an effect on her that she forgot all about me and off she went to see her. Then I came out of the cellar and bid them all good-bye and went down into the City for Mobile was well fortified and we lived inside of the fortifications and could by no means get out without a pass. And I went down in the City for safety being there was so many people passing[21] and I would not be easily detected.

So I wandered about in the City from one house to another where I had any friends for bout a week or more all night. So one morning as I went out of my friend's house quite early as I was compel to go for fear that I might be caught there. so I seen an old stable and I seen nobody around. I went in it for safety until I could go somewhere else. But I had not been in there long before in come a rebel Policeman with a pistol in his hand, cocked, and his finger on the trigger, and the cap was shining. he run to me and put the pistol in my breast and ask me what was I doing in there and who did I belong to. I told him. So he took me by the back of the neck and by the seat of my pants and took me throug the streets in a most ridiculous manner to the Guard House and put me in there. Then he went at Ten oclock and notified my Master that he had caught his boy Wallace. so the old man came to the Guard House and called for his boy Wallace. So they brought me out and took me over to the whiping house. And the man that whiped asked my Master did he want me whiped, the old man said yes. So they had my pants off and tied me in the ropes that was tied up against the wall So that the criminal might be clear the floor and could not look behind himself to see what they were doing to him.

So they tied the rope around my legs and arms. And they had a strap there about two or three leathers thick. and they hit me thirty lashes with that strap. They would hit me Ten licks and then let me cool off a while and then give me Ten more. Now when the man that whiped me had hit me Twenty licks he asked my Master would that do. he said, give him ten more, and every lick took the skin off. That was the worst whiping that I ever had. Then they took me down out of the ropes. and the old man said now see if you can go home. Now as I said Mobile was well fortified. The way that I had to get out of Mobile went right through the breast work. And that was only cut large enough for the road to go through and a guard stood in the way to keep any one from passing. Now where I lived was by way of this place. So instead of going home, I thought I would by the help of the lord make my second attempt[22] to make my escape.

Now there was about Twelve Thousand Confederate soldiers in Mobile. So I took courage and went on until I came in sight of the confederates, and I prayed faithfully to the lord to remove me out of that city. Now when I seen the soldiers and how they were strewed all along the way I was going, and that I could not get to the breastwork without going through their camp, my heart failed; however there being a little skirt of wood there, I lay down and took a knap of sleep. When I awaked I prayed again and I was encouraged to go on. So I put my trust in the Lord and went on and let the result be what it would. and I went right through their camp and they one to another looked at me but said nothing and when I got to the gap that led through the breast work there was no guard there.

So I got through the breast work safe. Then I had to go about Three miles before I got to the woods, this space of ground was cleared for a battle ground but the road passed through this space. So I reached the woods safely before dark. Then I took a woods course toward Cedar Point Ala which is on Grants Pass, opposite Dauphine Island.[23] But I tell you that I had a hard time before I got there.

Now as I had asked the lord to remove me safe from their hands, I believed it would be done even as I had asked. In my way I had many snakes to contend with and that of the bad kind. some of them would never move until they got a strike at me. I was bare footed and therefore greatly exposed to their venom, but it was the lord that carried me through and not myself. I was troubled all the day long with these snakes but I saw nothing of them at night. I never seen so many snakes in my life before or since. it was in the swamps south of Mobile where I walked over water and wanted it, but I could not drink it on the account of so many snakes. I went four days without anything to eat but a few grapes off of a grape vine I found in the woods.

About this time I could see the warships of the Union lying way off in the pass and in Mobile Bay. But there was betwixt me and them a dreadful looking river, called Foul river and a rebel picket line to cross, and that was no easy thing to do. Now the looks of the river frightened me. So I waited a day or so to see if I could not get across some other way without swiming but I could not. so I come across a cantaloupe patch and I was so near starved out that I destroyed every thing that was fit for any thing in that Patch that day; and I saw a color[e]d man that night and asked him for something to eat and he give me so much that it made me sick. Now as I come so near being caught that night and day or so after that, I resolved that I would try to get from that side of the river; it was death to go back and it was death to stay there and freedom was before me; it could only be death to go forward if I was caught and freedom if I escaped.

However to make my story short I went down to the river very early in the morning though the river looked very frightful but I put my trust in the Lord and I walked in the river as though there was no river there. with my hat, pants, jacket all

on the grass laped in the river on either side and the water was
deep; as I walked in my feet went down. So I had to swim
without wading any. I heard big fish braking about in the water.
Now when I got very near the shore on the other side I give out,
and sink but I found no bottom. when I rose again I swum out
and went on my way feeling very delighted that I had got across
the river that I dreaded so much.

But I had not gone far before I seen that I was in the Rebel
Picket lines; the place that I had to pass through was an old
open Piny woods, and it had broom sage[24] in it just about awaist
high and that was all that I had to hide in. Now the Pickets
lines was arranged in this way. Two men would ride up one road
and two men would ride down the other and so on. these roads
was large enough for Cavalry to ride two abreast; the roads was
as near as I can judge about a hundred yards apart. Now the way
I had to cross these lines when I seen two men ride up I would
run across the road, and when I seen two men ride down the
road, I would lay down in the broom sage until they passed,
then I would run across that road and do likewise, and so on
until I got across the Pickets lines.

Then I thought I was safe but I was not, just as soon as I
came out in the opening where the bushes did not hide my head
I heard some one holloring after me, and when I looked across
the opening I seen a man on an elevated road on a horse with a
gun in his hand waving it, I could see it shine. for fear it might
be a rifle, and he might shoot me I layed down and crawled
away from that place for that whole plain was a salt meadow
and there was water continually on the ground. So I could not
lay down except I bent the bushes down and lay upon them, so I
bent the bushes down in a thick place where they would keep

me off of the wet ground and lay there all day until dark. Then I went down to Cedar Point opposit Dauphine Island, and when I got there I was disappointed again. I saw that Fort Powell lay on the farther side of Grants pass, next to Dauphine Island, near the channel. That Fort was built by the Confederates some time before to prevent the Union gun boats from coming in to Mobile Bay through the Pass; but when they saw that they could not hold it, they put fire to the Magazene and escaped by night to Cedar Point and the Fort blow up destroying everything. Then the Federalls took the Fort and mounted one large gun to prevent an attck by the rebels from Cedar Point and also a gun boat to protect the pass by night.

Now there was another channell betwixt Fort Powell and Grants Pass, but it was blockaded. However as I told you when I got to Cedar point I was disappointed. I had to lay on Cedar point from one Saturday until the next, before I could get away, and I had no fresh water to drink in that time, and nothing to eat but what I could pick up where the soldiers camped months pasted. I had a hiding place in the ditch of the old Fortifications where the young reeds grow up very thick. The Rebels came down on that point once or twice every day to see if any yankee straglers had landed there; I came near being caught several times.

Once I became so impatient seeing the free country in view and I still in the slave country, I took me a long pole, and got on a log for the tide brought old logs ashore every day. I got one that would hold my weight and not sink. I thought by this way I could pole my way over. But to my surprise I soon got where I could not get bottom with my long pole; and the tide was taking me right up the bay, toward the Rebels. Now my log was very long and my mind led me to go to the upper end and I might find bottom, so I did. My log was very large and I could run on it just the same as you could on any large flat log. so I run to the upper end of the log and I found bottom just enough to arrest the speed of the log but I could not get the log to the shore again, for I was afraid that the rebels would come down before I got to shore, for it was not light when I left the shore, but the sun was quite high when I reached the land again. there was a long wharf there and when I got my log manageable I brought it up by the side of this wharf and got upon the wharf and run into my den. the rebels was late coming down that morning and I had not much more than got in my den before down they came.

Now they had a high spie house built to spie from and the mosquitos being quite bad there, I slept up there mostly anights; so one morning I slept quite late when I awaked I hastened down for fear they would come upon me and I had not been in my den long before they come down; and another morning I awaked while it was yet dark and for fear I would sleep late again, I came down. And just about light I was aroused in my den by the clinking of a rebel sabre climbing up into that spie house, just see how that the Lord delivered me from their hand.

Now on Friday I became very much troubled in mind, though I had got that far, yet I was not safe, and it was death if I was caught there; the Lord had brought me that far, I believed he would carryd me farther. So I prayed faithfully on Friday night that the Lord would remove me from that place. Now the tide commenced to go out every night and continued to go until the next day when it would return, bringing in my estimation the same logs and trash with it, though they might not been. After I had prayed for deliverance .from that place, I lay down to sleep, believeing it would be just as I had asked. The next morning I was awakened a great while before day by a voice singing within me; so I got up and came down out of the spie house, and it seemed as though some one invisible led me and it took me down by the side of the water; and where I could walk then I could not walk that day at twelve Oclock for the tide would be in.

Now when I got down there I seen a little boat very small indeed though the tide was going out. it stood like it was held by an invisible hand; so I got in the little boat and it held me, meantime I got a piece of board for an oar and started out from Cedar point to row my way over to Fort Powell. I got along very well until I got to a little sand Island between cedar point and fort Powell. Now when the tide was high you could not scarcely see this little Island, but when you got there it was large enough to hide behind. Now I had to hurry to get to this little Island, because my boat had diped water crossing the first channel. Moreover the Rebel Pickets could have reached me with their muskets if they had come down early. it was very squally that morning but I reached that little sand bar or Island before I got into a squall. When I got there I looked back from whence I came with delight, to think that I had come that far; and had not received a rebel ball.

I bailed out my little boat, and then set my coast [course] for fort Powell. Now I only had about a quarter of a mile to go before I was at the fort; they had a large cannon pointed direct toward me. Every now and then I could see them rise up on fort

Powell and then disappear, to see who I was. they could see the rebels at a distance spieing me. and the yankees could not understand it because I come from their country.

Now I had many thoughts in my mind about the appearance of things not withstanding. I believed the yankees to be my friends. so I started out for fort powell with the expectation of soon being in the free States. But when I had got about twenty five yards from this sand bar, I heard a voice according to my understanding say to me you cannot get up there your self, they will have to come out and get you. Now I began to doubt this, supposing this was my own imagination, and said what will hinder me from getting up there. In the meantime I looked up the bay and I seen the water like a hill coming with a white cap upon it. then I commence to see the trouble and while I was amusing in my mind what to do, the storm struck me. it carried my boat some distance unmanageable by me. I got into a trough and came near being swamped.

I understood a little about the water, but I could not done anything if the Lord had not helped me. So I managed to get my boat straight so as to ride the waves. in the midst of my struggle, I heard the crash of oars and behold there was eight Yankees in a boat. they run up by the side of me and said jump into this boat boy and in the midst of the motion of their oars I jumped into their boat and my boat turned bottom upward just as soon as I had vacated it. And they were struck with silence.

Then one of them said take that boat and bring it up to to show what a trifleing thing he came over in. The sea was high and they could not stop. They had to do all very quick. Then they turned their boat to go back to the fort again. All of this time, the Rebels was spieing me and them from their side for it was not no more than about two miles. The federal soldiers asked me where did I come from. I told them. the captain said he did not intend to come out to get me at first, but when he seen me in the storm, he got his men and came immediately to my relief.

And after pulling about a half an hour, we arrived at the Fort. After we arrived at the fort, the captain asked me no more questions at all. his men took me to their tents and give me attention in all that I wanted and I stayed on Fort Powell all night. The next morning I was up early and took a look at the rebels country with a thankful heart to think that I had made my escape with safety after such a long struggle; and had obtained that freedom which I desired so long. I Now dreaded the gun, and handcuffs and pistols no more. Nor the blewing of horns and the running of hounds; nor the threats of death from the rebel's authority. I could now speak my opinion to men of all grades and colors, and no one to question my right to speak.

So about eight oclock that morning I was ordered by the Captain, whose name was Captain Parker to be carried to Fort Gains [Gaines], to the General headquarters to be presented to Gen Granger[25] who was then commanding the department of the Gulf. he receivd me very courteously indeed and asked me the condition of Mobile and how the rebels stood there and I told him; he then told me as I had adventured through so many dangers to get there that he would give my my choice to enlist or to hire my self to some Officer. So I thought as I was so well known, that is among the rebels, in case I should be captured for there was much fighting to be done yet, I thought it best to hire myself to an officer which I did.

Now the officer's name was Julius Turner[26], and he was a gentleman and a friend to me; though we went in many dangers on one expedition, but we arrived back to the department of the gulf safe. I was his cook until Mobile fell then we were ordered to New Orleans from there to Natcheze Miss. and from there to Vicksburg Miss. And there they made out their muster roll and from there went to Baltimore Md. and were mustered out of service; and that was the way I got to Baltimore, Maryland which was in two states of where I was sold from. Here I heard from my people and seen them again, from which I had been absent from 1860 to 1868. So reader, I hope I have not maden[ed] you. so this wound up my fifth and last runaway.

My Dear reader, I have finished my book of adventures and struggles for freedom hoping you have approved of it. Don't take it for a novel, nor a fable, but a reality of facts. Oh that I may when done with this toilsome world, Even with three times the difficulties and persecutions that I met with in obtaining my temporal freedom, by God's assistance reach that Blistful abode, and triumph over the enemies of my soul at last. That will be a day of joy to me, which no tongue can express, for I will then be free indeed. Moreover my book is to show the goodness of God to me for his son's sake. when I prayed to him for my soul's freedom, he for Christ's sake freed my soul from the gall of bitterness and the bond of iniquity and he will not deliver me only but every one that believeth on him, and every one that trusteth in him. My reader I will now leave my book to your judgement. The End.

APPENDIX

John Washington's Eulogy to his infant son, John, May 10, 1865, reprinted by permission of Julian Houston and the Massachusetts Historical Society, Boston. The fate of children is a common theme in slave narratives, but such a separate eulogy in the writings of ex-slaves is extraordinarily rare.

The Death of Our Little Johnnie

Who Died March 2nd 1865 at 1:20 P.M.
Aged 8 Months and 25 Days

Who has not lost some dear little one, who used to be the joy of the household? But though the heart was then filled with sorrow, as if it would breake, and all the world looked lonely and sad.

Yet there still is consolation: For christ says, "Suffer little children to come unto me and forbid them not, for of such is the Kingdom of Heaven."[1]

After being racked with pains and wasteing fevers untill the little voice is scarcely audiable. The dear little bright eyes is closed never more to look up to us, his little hands lay by his side motionless, and he cannot talk to tell you of his necessities. When you can only think for the sufferer, and view his agony without any power to help, what a condition for the parents to be placed in!

But the messenger death comes and relieves the little sufferer from all earthly pain, and the battle thus just began is won; and the victory complete, and white winged angles waft its spirit upward to the skies. Where there is no sin and there is no weeping there; "for Christ has said of such is the Kingdom of Heaven."

"THE LITTLE GRAVE"

The winter is over with its cold bleak winds. The snow has melted before the warm bright sun of May, and over the little grave where the wind used to chant her mournful "dirges" now the sweet perfumes of spring flowers, and a thousand perfumes are wafted over the little grave, while the little spires of green grass point upward to God, the giver of every good and perfect gift in heaven. Where the weary cease trubling, and the mournors are at rest.

MAY 10, 1865
JMW

EPILOGUE

THE DISCOVERY OF DESCENDENTS

This book owes its existence to the remarkable "discovery" of these two rare slave-narrative manuscripts. As one of my primary objectives in writing the book was to uncover and recount the post-emancipation stories of each author, it was only natural that we hoped to find living descendents of John Washington or Wallace Turnage. Wallace Turnage's three surviving children died childless; hence there could be no direct descendents. But due to the diligence and skill of my research assistant, Christine McKay, we knew it was at least possible to find Washington's living relatives. The process involved detective work, the power of the Internet, the persistence to keep trying new combinations in the jigsaw puzzle of genealogy and modern American geographical mobility, and finally, some old-fashioned luck.

The result was the heartwarming discovery of Ruth Washington, Barbara Hinksman, and Maureen Ramos: the granddaughter, great-granddaughter, and great-great-granddaughter of John Washington. Few experiences in my life as a research historian have given me as much appreciation for the humane character of our craft as my encounter with these extraordinary women of strength, faith, and great intelligence; they have embraced the story of their ancestor, about whom they knew very little, with zeal and grace, and their response is exemplary of the difficult but rewarding process by which early twenty-first-century African Americans can find, understand, and learn from their enslaved ancestors.

I confess that at one point in 2007 I urged Chris McKay to give up the search. It appeared to only turn up false starts and dead ends. The book was otherwise ready for the publisher. But the ingenious Chris pushed on. We knew that John Washington had five surviving sons: William, James, John Jr., Charles, and Benjamin. Charles and James had no children; William, who moved to Chicago, had one son, Earl, who left two stepchildren, but no natural children; Benjamin had one daughter, Evelyn, about whom I have written earlier in the book and who kept possession of the narrative manuscript; and Evelyn had no children. John Jr. was our only hope. He lived in Jersey City, owned his own home, and worked for many years in a meatpacking factory. John Jr. left three surviving daughters—Priscilla, Francenia, and Ruth—and they were John Washington's only grandchildren. John Jr.'s 1936 obituary gave Francenia's married name as Hinksman and Ruth's as Rhetta, but this is about all we knew as *A Slave No More* was about to go to press.

Forays into the Jersey City Public Library turned up some information: Priscilla's married name had been Martin, all three sisters had attended college, and all had been teachers in the Atlantic City area. Chris continued to run various combinations of all these names through search engines, and our lucky break came from a May, 2007 obituary from Atlantic City. The deceased was clearly a relative (an in-law of Priscilla's), and the notice provided survivor names, including Hinksman and Martin—and a "Ruth Washington, Tampa, Florida." Chris called me with the good news. We started calling Tampa.

Using public records, Chris had also located Barbara Hinksman, Priscilla's daughter, who confirmed that she was related to John Washington. Barbara in turn gave Chris her Aunt Ruth's telephone number. Chris made the first contact with Ruth, the eighty-nine-year-old granddaughter of our slave-narrative author. She was delightful and completely cooperative on the telephone. Then it was my turn to call Ruth. It can only be described as an out-of-body experience to call someone and tell her that you are about publish her grandfather's autobiography about his experience as a slave.

Ruth lives in a trailer-park retirement village in Tampa; she is a retired teacher but still drives her own car and works as a teacher's aide several days per week. Ruth confirmed that she is Washington's granddaughter, but explained that she knew very little about him or her grandmother, Annie. She had never met them. John Sr. died the year Ruth was born, in 1918. Tellingly, her father never talked about the past. Her parents, she said, kept a photograph on the wall of her grandparents, but never discussed their story, not even where they were buried. After a divorce, Ruth had taken back her maiden name. Her one son had died many years ago, as had her sisters. Ruth remarked that she did not know why she was still alive, but that with this news about her grandfather's narrative, we had "given her back her family" and helped her "know why she was still here."

Immediately after our telephone conversations, in late October 2007, I flew to Tampa to meet Ruth and her niece, Barbara. On an unforgettable Saturday night, Ruth, Barbara, and Maureen (Barbara's daughter) met me at the airport. With Ruth at the wheel, they drove me to her favorite restaurant at a nearby mall, the Cheesecake Factory. There, for many hours, we sat and talked, shared some documents and photographs, and otherwise explored the past of this extraordinary American family. They opened their hearts to me, a stranger. Before the night was over, Ruth showed me her family Bible and one photograph that, she said, she understood to be her grandparents' gravesite. I gulped, and showed her my own photograph of the same gravestone, one that appears in this book. Ruth had no idea where it was located and I had the honor of informing her that her grandparents' resting place was in Cohasset, Massachusetts.

I had mailed them copies of the book and they were already well informed of John Washington's story. Ruth easily remembered her Uncle Benji, with his shiny bald head. Most extraordinary, however, was the glaring fact that these well-educated people had been cut off from their own family past. They knew next to nothing about their slave ancestors, not even that their people had come from Virginia. This was doubly ironic because Ruth and her sisters had all attended Hampton University in Virginia, the black college created during Reconstruction. Their father, according to Ruth, raised her and her sisters to be very proper young ladies, and he struggled to make sure they went to college in the Depression years of the 1930s. Ruth told a moving story about her father's death, which occurred her freshman year at Hampton. He had been a formal man, she said, and did not wish to discuss the past, even when asked. That reaction was all too typical in the early twentieth-century among many African-American families struggling to establish stable, working-class lives and respectability for themselves. As I have written elsewhere: facing a past of slavery, telling the story of its legacies, feeling the markers of slavery on one's identity into the generation that followed emancipation was never easy in an America that constantly stressed "progress" and segregated blacks in so many aspects of life and work. That Ruth had been cut off from her grandparents' story in the Jim Crow era did not surprise me as a historian; it moved and troubled me as a human being and a citizen.

My relationship with Ruth, Barbara, and Maureen grew. As part of my book tour in November 2007, the National Park Service in Fredericksburg helped me plan a very special event in Ruth's honor. I flew Ruth and Maureen up to Virginia and we staged a one-day celebration of John Washington's story in Fredericksburg. Ruth was treated like a visiting celebrity, and a reporter from the local newspaper followed her around for interviews. We visited the sites of John Washington's youth, including the room in the Farmer's Bank where he slept and said his goodbyes to his mother. Most movingly, we walked down to the spot on the Rappahannock River where John had crossed to freedom on April 18, 1862. There, John Hennessey of the Park Service told the story aloud, and I added some details as well. But Ruth, with glistening eyes, made the story real as she marveled at the courage of her grandfather.

That evening, at the old Baptist Church in Fredericksburg, the city and the Park Service staged a special event attended by some one hundred and fifty to two hundred people. Hennessey and others delivered a dramatic reading from part of Washington's narrative, and a choir sang appropriate songs. I gave a talk about the book and its roots in Fredericksburg. Then I introduced Ruth, who stole the show. Beautifully attired in her Sunday best, flowers on her lapel, she stepped to the pulpit, all four foot ten of her. She took from her pocket a script written on notebook paper. In her short, fifteen-minute address, she thanked the people of Fredericksburg, discussed what she had learned from the book, marveled at all of her new friends, and spoke movingly of her religious faith. She also stressed how important it was for her to learn how literate her grandfather had been. Then she talked about how meaningful it was for her to discover and learn of her slave ancestors; that it was no stigma, but rather a source of great pride that her people had survived and escaped from slavery in the Civil War. With her strong voice, Ruth became the embodiment of how the past is a living, breathing thing, a reminder that we must never say never as we search for it and face what we find. Ruth had the audience laughing and weeping and, as she finished, on their feet cheering. An eighty-nine-year-old African American woman of short stature had told us all what it means to know where we are from, to experience the bracing joy of knowing how to fill the mysterious holes in one's very identity. At the book signing that followed, Ruth signed the books with me; it had become her story as well as her grandfather's. For a night, Ruth was a rock star in a Virginia town that has not easily embraced its past in relation to slavery and the deepest meanings of the Civil War.

It is never easy to know just what bloodlines mean to people. Some descendents of former slaves—or of anyone else, for that matter—may have had quite different reactions from that of Ruth, Barbara, and Maureen. But these remarkable women, who are now a kind of extended family for me, demonstrate that whatever bloodlines mean, they convey historical lines and paths to understanding who we are in time, in a universe that will erase our stories unless we remain vigilant to find, preserve, and tell them. Ruth Washington looks a great deal like her grandfather. In John's photographs he exudes a certain modesty, respectability, and pride in his family. He has a joy in his eyes, if not a bursting smile. Ruth, Barbara, and Maureen (teachers all—he would be so proud!) are smiling for him now.

DAVID W. BLIGHT
AUGUST, 2008

ACKNOWLEDGMENTS

A LEGION OF PEOPLE HELPED ME RESEARCH AND WRITE THIS
BOOK, but none more than Christine McKay. Chris is as talented a
researcher into obscure sources about the lives of ordinary people as
one will ever find. I would never have rounded out the post-slavery
lives of Washington and Turnage without Chris's help. She also
saved me from several errors of fact. Fellow historians take note: if
you cannot find something, call Chris. She will find it or know why
it can't be found.

My agent, Wendy Strothman, had much to do with conceiving
this book. I am most grateful for her skill, her readings, and her faith
in me. Dan O'Connor, also of the Strothman Agency, has provided
invaluable assistance in bringing attention to these two rare documents. Julian Houston, who owns the Washington narrative and papers, first put his confidence in Wendy and then in me, for which I
am grateful. Julian's mother, the late Alice J. Stuart, compiled considerable information on Washington's narrative and made writing
this book much easier than it might have been. Debra Mecky and
her staff at the Historical Society of the Town of Greenwich, especially archivist Ann Young, have been a joy to work with. We are all
grateful to Gladys Watts for calling the Historical Society in Greenwich to inform them of Turnage's narrative. I thank the staff of the
Massachusetts Historical Society in Boston for helping me so graciously on my several visits, especially Donald Yacovone and Carolle

Morini. David Wadsworth of the Cohasset, Massachusetts, Historical Society is a revelation and a walking local memory for his town. Gordon Poindexter of the Library of Virginia in Richmond gave enormous assistance in ferreting out information on the slave trader Hector Davis. Leslie Martin of the Chicago Historical Society helped me use the Hector Davis & Company Account Books.

John Hennessy, historian for the National Park Service in Fredericksburg, Virginia, and a distinguished Civil War historian, has been my angel of Maryes Heights. John gave me tours, much valuable information, and extraordinary assistance with photographs. I thank Alan Zirkle, who took special photographs for me in Fredericksburg. I also have an angel of the archives as well as of the local lore in Mobile, Alabama. Ann Dowdle Maddox took me on an unforgettable tour of the Fowl River estuary, Mobile Bay, Dauphin Island, and one fine country oyster and shrimp restaurant near where Turnage crossed the Fowl River. I am in debt to Katherine Mooney for directing me to Ann and for her uncommon sense of humor. Michael Fitzgerald was also a source of good advice on the history of Alabama and Mobile. The University of South Alabama Archives provided many ideas for illustrations. Archivist Barry McGhee of the Fredericksburg Circuit helped us locate some important John Washington Documents.

Bill Nelson was my cartographer who produced two splendid original maps for this book. I owe a tremendous debt to Pembroke Herbert of Picture Research Associates for helping me find many key photographs and drawings for the book. Pembroke is so good at what she does that she makes it seem effortless. Karin Beckett was an enduring source of advice, grace, moral support, and conversation about this book from the day I started. Shawn Alexander, William Andrews, James Basker, Klaus Benesch, Randall Burkett, James

Campbell, Clark Dougan, Henry Louis Gates, Jr., Jeffrey Ferguson, Eric Foner, James and Lois Horton, Waldo Martin, Deborah McDowell, Colin Palmer, Caryl Phillips, and John Stauffer all helped me think about how to write this book. Skip and Bill provided sage advice as only they can, and Jeff gave indispensable friendship. Louise Mirrer and the Board of Trustees of the New York Historical Society listened intently to a brief version of these two tales. Marni Sandweiss read chapters and gave me valuable encouragement and research tips. James Oakes, an expert on emancipation, also read chapters and offered astute suggestions.

I extend a very special thanks to my editor at Harcourt, Andrea Schulz, whose stewardship, diligence, and faith in this project have been astonishing. Michelle Blankenship graciously helped me think of ways to find readers. I did much of the final writing and some of the research while a fellow at the Dorothy and Lewis E. Cullman Center for Scholars and Writers at the New York Public Library. The NYPL is a national treasure. The Cullman Center is a scholar's dream, and as delightful a sanctuary as any writer could ever need. I thank the Cullman director Jean Strouse and her staff, Pamela Leo, Elizabeth Bradley, and Adriana Nova. For all the conversations I am in debt to Cullman fellows Muhammad Ali, Sharon Cameron, Will Eno, Clive Fisher, Farah Griffin, Maya Jasanoff, Carla Kaplan, Ben Katchor, James Miller, James Shapiro, Laurie Sheck, Nelson Smith, Jeff Talarigo, and Sean Wilentz. The staff of the New York Public Library helped me endlessly, especially Alice Hudson in the map collection and Thomas Lisanti in photographs.

At Yale University I thank my research assistant, Steve Prince, and numerous librarians. Chip Long and Lloyd Suttle of the Provost's office and Dean Jon Butler made many things possible. Pamela and John Blum and David and Toni Davis have made living with this book and in New Haven enjoyable. My colleagues at the Gilder Lehrman Center for the Study of Slavery, Resistance and Abolition have tolerated my stories, complaints, and requests for the more than two years this book has evolved. Dana Schaffer helped me finish the book in more ways than can be counted. Melissa McGrath and Angela Kaiser deserve applause, and Tom and Debra Thurston extended me crucial assistance in transcribing and scanning each narrative. I also thank Tom for being intellectually simpatico and for his advice. Rob Forbes was present at the creation. I am further grateful to all my colleagues in the Gilder Lehrman working group at Yale on "Slavery in the Artistic and Historical Imagination" for critically reading a chapter of this book. For their grace, inspiration, and friendship, I extend special thanks to Washington descendants Ruth Washington, Barbara Hinksman, and Maureen Ramos. Finally, a special thanks to Marsha Andrews for her research assistance, keen interest in language, appreciation for my prose, and above all companionship.

NOTES

PROLOGUE

1. *Douglass Monthly,* May 1861.

2. W. E. B. Du Bois, *Black Reconstruction in America, 1860–1880* (1935 rpr. New York: Atheneum, 1962), 125; Frederick Douglass, *My Bondage and My Freedom* (1855 rpr. New York: Collier Books, 1969), 272.

3. "N.L.J. to William Still," April 16, 1859, in Carter G. Woodson, ed., *The Mind of the Negro as Reflected in Letters Written During the Crisis, 1800–1860* (Washington, D.C.: Association for the Study of Negro Life and History, 1926), 563.

4. Hamilton Holt, ed., *The Life Stories of Undistinguished Americans, As Told by Themselves* (1906 rpr. New York: Routledge, 1990), Werner Sollors, ed. See Sollors's foreword, "From the Bottom Up," xi–xxviii.

5. "The Life Story of a Negro Peon," in ibid., 122–23; "The Life Story of a Southern Colored Woman," in ibid., 218. These two pieces first appeared respectively in *The Independent,* February 25, 1904 and March 17, 1904. Both are called "autobiography" by the editors but were in effect interviews.

6. Richard Wright, *Twelve Million Black Voices: A Folk History of the Negro in the United States* (New York: Viking Press, 1941), 46–47; Frederick Douglass, "Speech at the Thirty-third Anniversary of the Jerry Rescue," 1884, Douglass Papers, Library of Congress (LC), reel 16.

7. See Harriet E. Wilson, *Our Nig, or Sketches from the Life of a Free Black, in a Two-Story White House, North: Showing that Slavery's Shadows Fall Even There, by "Our Nig",* Henry Louis Gates, Jr., ed. (1859 rpr. New York: Vintage, 2002); Hannah Crafts, *The Bondwoman's Narrative,* Henry Louis Gates, Jr., ed. (New York: Warner Books, 2002); Julia C. Collins, *The Curse of Caste; Or the Slave Bride: A Rediscovered African American Novel,* William L. Andrew

and Mitch Kachun, eds. (New York: Oxford University Press, 2006). On Crafts, also see essay that exposed the text, Henry Louis Gates, Jr., "Borrowing Privileges," *New York Times Book Review,* June 2, 2002.

8. Gates, ed., *The Bondwoman's Narrative,* xxxii. *The Bondwoman's Narrative* has received some very positive as well as some skeptical reviews and responses. A series of essays on the text are collected in Henry Louis Gates, Jr. and Hollis Robbins, eds., *In Search of Hannah Crafts: Critical Essays on The Bondwoman's Narrative* (New York: Basic Civitas Books, 2004). The true identity of Hannah Crafts has not yet been determined.

9. For the most complete and authoritative bibliography of antebellum and postbellum slave narratives, see *docsouth.unc.edu/neh/bibliography,* edited and compiled by William L. Andrews. For 1745–1865, Andrews includes 102 published titles, but approximately a third were second editions, reprints, or translations of older narratives. For 1870–1920s, Andrews counts 72 total titles, but approximately a third of those are again reprints of earlier editions. It is in this second grouping of postbellum slave narratives where Turnage and Washington belong. On antebellum slave narratives as a genre, see William L. Andrews, *To Tell a Free Story: The First Century of Afro-American Autobiography, 1760–1865* (Urbana: University of Illinois Press, 1988); and Charles T. Davis and Henry Louis Gates, Jr., eds., *The Slave's Narrative* (New York: Oxford University Press, 1985).

10. On structure and conventions in slave narratives, see James Olney, "'I Was Born': Slave Narratives, Their Status as Autobiography and as Literature"; and Robert Stepto, "I Rose and Found My Voice: Narration, Authentication, and Authorial Control in Four Slave Narratives," in Davis and Gates, eds., *The Slave's Narrative,* 148–74, 225–41.

11. William L. Andrews, "The Representation of Slavery and the Rise of Afro-American Literary Realism, 1865–1920," in Deborah E. McDowell and Arnold Rampersad, eds., *Slavery and the Literary Imagination* (Baltimore: Johns Hopkins University Press, 1989), 62–80. For the "tomb" and "school" metaphors, see Douglass, *Narrative of the Life of Frederick Douglass, An American Slave* (1845 rpr. Boston: Bedford Books, 2003), David W. Blight, ed., 89; and Booker T. Washington, *Up From Slavery* (1901 rpr. Boston: Bedford

Books, 2004), Fitzhugh Brundage, ed., 47. On postbellum narratives written especially by black Union army veterans, see David W. Blight, *Race and Reunion: The Civil War in American Memory* (Cambridge, Mass.: Harvard University Press, 2001), 195–98, 437–38. There are at least ten to a dozen such narratives by veterans and most conform to this mode of an ascension story about spiritual evolution and success. See Robert Anderson, *From Slavery to Affluence: Memoirs of Robert Anderson, Ex-Slave* (Hemingford, Nebr.: Hemingford Ledger, 1927); and Elijah Preston Marrs, *Life and History of Rev. Elijah P. Marrs* (Louisville, Ky.: Bradley and Gilbert, 1885).

12. John Washington, "Memorys of the Past," 15; "Journal of Wallace Turnage," introduction. One recently published text that merits some comparison to Turnage and Washington is William B. Gould IV, *Diary of a Contraband: The Civil War Passage of a Black Sailor* (Stanford, Calif.: Stanford University Press, 2002). William B. Gould IV is the great-grandson of William Gould, a slave who escaped in a stolen boat on a river during the war and served in the Union navy. On September 22, 1862, Gould, a twenty-four-year-old skilled, literate, urban slave in Wilmington, North Carolina, commandeered a boat along with seven other young black men and rowed it twenty-eight miles out the Cape Fear River to the Atlantic Ocean, where they were picked up and ultimately freed by the *U.S.S. Cambridge*, a ship patrolling as part of the North Atlantic Blockading Squadron. During his three years' service until 1865, Gould kept a diary in which his entries are brief, disciplined, and almost daily for long periods of time as he sails all over the Atlantic. The document remained in an attic in Dedham, Massachusetts, where the Civil War veteran settled after the war, until it was discovered by his grandson in 1958. Like Washington, Gould's liberation was in part the result of his urban residence, skills obtained from hiring out, and opportunity for movement; like Turnage, he rowed his way to the Union navy. Like both, he wrote down some of that experience in a raw form for his family to discover. Unlike Turnage and Washington, Gould wrote a sailor's log rather than a self-conscious autobiographical story of escape.

13. On autobiography as an art form, see James Olney, *Metaphors of Self: The Meaning of Autobiography* (Princeton, N.J.: Princeton University Press,

1972), 264–74. "Journal of Wallace Turnage," introduction; "Memorys of the Past," 15–25; Jerzy Kosinsky, *The Painted Bird* (1965 rpr. New York: Grove Press, 1995), 234–35.

Chapter One: The Rappahannock River

1. "Memorys of the Past," 15–16.

2. On the mixed-race population in Virginia and that state's struggle to codify who was and was not black or white, see Joshua D. Rothman, *Notorious in the Neighborhood: Sex and Families Across the Color Line in Virginia, 1787–1861* (Chapel Hill: University of North Carolina Press, 2003).

3. Obituary for Thomas R. Ware, *Virginia Herald*, September 18, 1820, Central Rappahannock Public Library, Fredericksburg, Va.

4. Thomas R. Ware's Will, January 13, 1820, and estate inventory of Catherine Ware, April 5, 1825, Thomas R. Ware Papers, Fredericksburg Circuit Court, Richmond, Va. Molly's whipping payment is recorded in the 1829 Ware estate accounting, Will Book C, p. 117, June 29, 1829, Ware Papers, Fredericksburg Circuit Court. "Boarders Wanted," a notice placed by Catherine Ware, *Virginia Herald*, March 2, 1825, Central Rappahannock Public Library. Grandmother Molly is likely to have died in Washington, D.C., sometime in the 1860s. She does not appear in the 1870 D.C. census, and death certificates were not recorded in the District until 1874.

5. *Day Book* of Dr. James Carmichael, Fredericksburg, Va., James Carmichael Papers, University of Virginia Library, Charlottesville, Va., http://hsc. Virginia.edu/hs-library/historical/carmichael. Carmichael was a physician to the Ware slaves. On January 14, 1817, he reports a visit to a "Negro of Thomas Ware," and on February 23, 1817, he notes the vaccination of three Negroes for Ware, presumably Molly, Alice, and Sarah. For Sarah's "bright mulatto" complexion, see Certificate of Death for Sarah Tucker, May 3, 1880, Vital Records Division, Department of Health, District of Columbia. At some point before the Civil War she had married Thomas Tucker, and she is listed as a "widow" on her death certificate.

6. On Brown, see U.S. Census, 1840, Orange County, Va. Marriage notice for Catherine Ware and Francis Taliaferro, Esq., in *Virginia Herald*, October 24,

1832, Central Rappahannock Public Library. "Negroes Wanted," *Political Arena*, October 7, 1836, Central Rappahannock Public Library.

7. "Memorys of the Past," 16–24.

8. "Ten Dollars Reward," Fredericksburg *Political Arena*, February 19, 1841, in Central Rappahannock Public Library.

9. John Hope Franklin and Loren Schweninger, *Runaway Slaves: Rebels on the Plantation* (New York: Oxford University Press, 1999), 212, 224.

10. Harriet Jacobs, *Incidents in the Life of a Slave Girl, Written by Herself*, Jean Fagan Yellin, ed. (1861; rpt. Cambridge, Mass.: Harvard University Press, 1987), 35.

11. Ibid., 37, 56.

12. "Memorys of the Past," 25, 28.

13. Ibid., 25–26.

14. Frederick Douglass, *Narrative of the Life of Frederick Douglass, An American Slave, Written by Himself*, David W. Blight, ed. (1845 rpt. Boston: Bedford Books, 1993), 64; Richard C. Wade, *Slavery in the Cities: The South 1820–1860* (New York: Oxford University Press, 1964), 245; *De Bow's Review*, no. 29, 1860; John S. C. Abbott, *South and North; or Impressions Received During a Trip to Cuba and the South* (1860), in Wade, *Slavery in the Cities*, 246.

15. *New Orleans Daily Picayune*, March 8, 1856, in Wade, *Slavery in the Cities*, 261.

16. Ralph Ellison, "The World and the Jug," in Ellison, *Shadow and Act* (New York: Vintage, 1964), 116.

17. "Memorys of the Past," 29–30, 41. For Washington's membership in the African Baptist Church, see *www.shiloholdsite.org/members1854.htm*.

18. "Memorys of the Past," 31.

19. Ibid., 32–33.

20. Douglass, *Narrative*, 42; Charles Ball, *Fifty Years in Chains: Or the Life of An American Slave* (1837 rpt. New York: H. Dayton, 1858), 10–12; Josiah Henson, *The Autobiography of the Reverend Josiah Henson* (1849 rpt. Reading, Mass.: Addison-Wesley, 1969), 17–18.

21. "Memorys of the Past," 34–38. Washington also mentions his Sunday school teacher, Olive Hanson, "a most kind and gentle lady" (40), as another person who may have helped him with his literacy.

22. Ibid., 41–54.

23. Ibid., 62, 56; Douglass, *Narrative*, 83–84.

24. "Memorys of the Past," 61–62; Langston Hughes, "The Negro Speaks of Rivers," 1926, in Nathan Irvin Huggins, ed., *Voices of the Harlem Renaissance* (New York: Oxford University Press, 1995), 155.

25. U.S. Census, 1850, Spotsylvania County, Va. Annie's mother was Virginia Gordon, a free black woman. The 1860 Census lists Annie living with a free black family named Jackson, and working as a seamstress. "Memorys of the Past," 58–59; untitled manuscript fragments of courtship, approximately 25 pages, 1856, 1858, John Washington Papers, Massachusetts Historical Society (MHS).

26. Untitled manuscript, courtship letter and diary, Washington Papers (MHS).

27. "Memorys of the Past," 59; Wade, *Slavery in the Cities*, 117–20.

28. "Memorys of the Past," 63–64; Frances Anne Kemble, *Journal of a Residence on a Georgia Plantation in 1838–1839*, John A. Scott, ed. (New York: Knopf, 1961), 122–23, quoted in Jonathan D. Martin, *Divided Mastery: Slave Hiring in the American South* (Cambridge, Mass.: Harvard University Press, 2004), 115.

29. "Memorys of the Past," 64–65.

30. Ibid., 66; Bruce Jackson, *Wake Up Dead Man: Afro-American Work Songs from Texas Prisons* (Cambridge, Mass.: Harvard University Press, 1972), 30, quoted in Lawrence W. Levine, *Black Culture and Black Consciousness: Afro-American Folk Thought from Slavery to Freedom* (New York: Oxford University Press, 1977), 215.

31. "Memorys of the Past," 68–69. On self-hiring, see Martin, *Divided Mastery*, 161–87. One study has found that in 1860, of Virginia's total slave population, 246,981, hired slaves numbered approximately 25,000, and among that number, 2,500, or 10 percent, were self-hired. See Loren Schweninger, "The Underside of Slavery: The Internal Economy, Self-Hire, and Quasi-Freedom in Virginia, 1780–1865," *Slavery and Abolition* 12 (September 1991), 10.

32. "Memorys of the Past," 68. On President Jefferson Davis's creation of a "Confederate ideology" and justification for war as a struggle against Northern "tyranny" and for "self-government" and "liberty," see Paul D. Escott, *After Secession: Jefferson Davis and the Failure of Confederate Nationalism* (Baton Rouge: Louisiana State University Press, 1978), 36–53.

33. Letter, John Washington to Annie Gordon, Richmond, Va., October 27, 1861, John Washington Papers (MHS).

34. "Memorys of the Past," 70–71.

35. Ibid., 72. For a scholarly demolition of the "black Confederate" claims of modern enthusiasts, see Bruce Levine, *Confederate Emancipation: Southern Plans to Free and Arm Slaves During the Civil War* (New York: Oxford University Press, 2006).

36. "Memorys of the Past," 72–75.

37. The casualties at Shiloh were 13,047 Union and 10,694 Confederate.

38. Maj. Gen. Irvin McDowell to E. M. Stanton, Sec. of War, Aquia, Va., April 18, 1862; report of Brig. Gen. Christopher C. Augur, camp opposite Fredericksburg, Va., April 18, 1862; W. W. D. to Col. J. E. Johnson, near Port Royal, Caroline County, Va., April 20, 1862, *The War of the Rebellion: A Compilation of the Official Records of the Union and Confederate States*, series I, vol. XII (Washington, D.C.: Government Printing Office, 1885), 427–30, 437 (hereafter cited as *Official Records*).

39. "Memorys of the Past," 77–80.

40. Ibid., 80.

41. Ibid., 81, 83–86. Note: Washington places his map in the narrative right after he had returned the keys to the Shakespeare's proprietor's wife, as if to say, "I am free of all of slavery's demands; let me now show you my path to liberation."

42. Ibid., 87.

43. Ibid., 88–90. The 30th and 21st New York regiments were in the same brigade.

44. Ibid., 91–93.

45. J. Harrison Mills, *Chronicles of the Twenty-first Regiment, New York State Volunteers* (Buffalo, N.Y.: 21st Regiment Veteran Association of Buffalo, 1887), 166–67. Microfilm copy, Sterling Library, Yale University, New Haven, Conn.

46. "Memorys of the Past," 90, 94–96.

47. Ibid., 96–99. Washington claims some thirty Fredericksburg Confederates were arrested and that Reuben Thom, a close family friend of the Wares, Washington's former owners, and a man John had known well since his youth, was one of them. Those arrested included only nineteen residents of the city,

taken in two waves: seven on July 22, which is when Washington participated as a guide, and twelve on August 13, after he had gone north again. And, he mistakenly claims Thom as one of the hostages. Thom was eighty-two years old, a revered postmaster, and was not arrested. Washington may have harbored animosities toward Thom that led to his claim. I am grateful to John Hennessy of the National Park Service in Fredericksburg for this information.

48. Ibid., 99–100.

49. Ibid., 101–05. On Mosby, see Mark M. Boatner, *The Civil War Dictionary* (New York: David McKay Co., 1959), 571. On Mosby in history and memory, see Paul Ashdown and Edward Caudill, *The Mosby Myth: A Confederate Hero in Life and Legend* (Wilmington, Del.: SR Books, Scholarly Resources, 2002); and David W. Blight, *Race and Reunion: The Civil War in American Memory* (Cambridge, Mass.: Harvard University Press, 2001), 293, 297–99.

50. "Memorys of the Past," 107–10.

51. Ibid., 111.

52. Ibid., 112–13. On the battles of the Seven Days, see James M. McPherson, *Battle Cry of Freedom: The Civil War Era* (New York: Oxford University Press, 1988), 464–71, and for "thrust and parry," 526. The activities and locations (especially various cavalry expeditions to determine Confederate troop deployment) of King's division can be followed closely by various dispatches in *Official Records,* series I, vol. 7, part III, 502, 514, 548–49. The latter helps confirm the remarkable accuracy (for at least dates and locations) of much of Washington's memory of these events. In Gen. John Pope to Gen. Rufus King, August 8, 1862, Pope orders King to march his division as soon as possible toward Culpeper, Virginia.

53. "Memorys of the Past," 112–19.

54. Ibid., 120–22. For a detailed account of the campaign that resulted in the battle of Second Manassas, see John J. Hennessy, *Return to Bull Run: The Campaign and Battle of Second Manassas* (New York: Simon & Schuster, 1993), 1–201. The Orange and Alexandria Railroad "ground to a halt" sometime in mid-August, writes Hennessy (39), due to bad management and military inefficiency, so Washington managed his departure just before this temporary cessation of rail traffic.

55. Hennessey, *Return to Bull Run*, 168–69, 170–93, 198. Pope's headquarters were still at Cedar Mountain on August 18, but by the 19th, his dispatches emanated from Culpeper. See *Official Records*, series I, vol. 7, part III, 592, 601.

56. "Address on Colonization to a Deputation of Negroes," August 14, 1862, in Roy P. Basler, ed., *The Collected Works of Abraham Lincoln* (New Brunswick, N.J.: Rutgers University Press, 1953), vol. V, 372. The delegation was led by Edward M. Thomas, the president of the Anglo-African Institute for the Encouragement of Industry and Art in Washington, D.C. The other four members were John F. Cook, John T. Costin, Cornelius Clark, and Benjamin McCoy. See also Michael P. Johnson, ed., *Abraham Lincoln, Slavery, and the Civil War: Selected Writings and Speeches* (Boston: Bedford Books, 2001), 200–04.

57. "Address on Colonization to a Deputation of Negroes," 371–75. For an interpretation of the address sympathetic to Lincoln's "racial candor," and arguing that it should be seen in the context of conditioning public opinion for emancipation, see Allen C. Guelzo, *Lincoln's Emancipation Proclamation: The End of Slavery in America* (New York: Simon & Schuster, 2004), 139–44. For a more critical interpretation, and one that also accounts for the outraged character of the black response, see David W. Blight, *Frederick Douglass' Civil War: Keeping Faith in Jubilee* (Baton Rouge: Louisiana State University Press, 1989), 137–40.

58. "Address on Colonization to a Deputation of Negroes," 375.

59. "Memorys of the Past," 122–29.

60. Ibid., 131–32.

61. Ibid., 133–34; Hennessey, *Return to Bull Run*, 439–72.

62. "Memorys of the Past," 134.

CHAPTER TWO: MOBILE BAY

1. Courtney Hart death certificate, City of New York #28553, October 7, 1898, lists her age at sixty-seven. For Wallace's identification of Sylvester as his father, see marriage certificate #2794, Health Department of the City of New York, May 10, 1875, and Freedmen's National Bank accounts #2090 and 5187. On forced interracial unions between white masters and slave women, see

numerous essays in Martha Hodes, ed., *Sex, Love, Race: Crossing Boundaries in North American History* (New York: NYU Press, 1999), especially Peter W. Bardaglio, "'Shameful Matches': The Regulation of Interracial Sex and Marriage in the South before 1900," 112–40, and Sharon Block, "Lines of Color, Sex and Service: Comparative Sexual Coercion in Early America," 141–63. See also Deborah Gray White, *Ar'n't I a Woman?: Female Slaves in the Plantation South* (New York: Norton, 1985); and Emily West, *Chains of Love: Slave Couples in Antebellum South Carolina* (Urbana: University of Illinois Press, 2004).

2. Wallace Turnage death certificate, #8060, State of New Jersey Bureau of Vital Statistics, October 5, 1916; and marriage certificate, Wallace Turnage, #13,096, Health Department of the City of New York, November 19, 1889. U.S. Census, 1830; U.S. Census, 1860.

3. "The Journal of Wallace Turnage," introduction.

4. Frederick Bancroft, *Slave Trading in the Old South* (Baltimore: J. H. Furst Co., 1931), 94–99; Dickinson, Hill, & Co. circular, quoted in Michael Tadman, *Speculators and Slaves: Masters, Traders, and Slaves in the Old South* (Madison: University of Wisconsin Press, 1989), 61. On the slave trade as layers of "fantasy" for owners seeking to fulfill their many needs, economic and psychological, see Walter Johnson, *Soul by Soul: Life Inside the Antebellum Slave Market* (Cambridge, Mass.: Harvard University Press, 1999), 78–93, 112–115.

5. *Hector Davis & Company Account Books, 1857–1864,* 84, 98, Chicago Historical Society Archives, Chicago, Ill.

6. *Business Directory of 1859,* Richmond, compiled by W. Ferslew and published by George M. West, in Works Progress Administration Historical Inventory, No. 231; "Slave Market of Hector Davis," June 11, 1937, copy in Library of Virginia, Richmond; Bancroft, *Slave Trading in the Old South,* 95–96.

7. "Journal of Wallace Turnage," 2; Tadman, *Speculators and Slaves,* 47–82; Johnson, *Soul by Soul,* 84–88, 135–161. Johnson refers to the "grammar of economic speculation" (84) to characterize the language through which this commerce was conducted. William Wells Brown, *The Narrative of William W. Brown, a Fugitive Slave* (rpr. 1848; Reading, Mass.: Addison-Wesley, 1969), 17–18.

8. "Journal of Wallace Turnage," 3–4.

9. Johnson, *Soul by Soul*, 4–8; Tadman, *Speculators and Slaves*, 130–31.

10. Johnson, *Soul by Soul*, 6. Also see J. B. Pritchett, "The Domestic United States Slave Trade: New Evidence," *Journal of Interdisciplinary History*, 21 (1991), 471–75; and Todd L. Savitt, "Slave Life Insurance in Virginia and North Carolina," *Journal of Southern History*, 43 (1977), 583–600. Also see the collection of essays, Walter Johnson, ed., *The Chattel Principle: Internal Slave Trades in the Americas* (New Haven, Conn.: Yale University Press, 2004).

11. "Journal of Wallace Turnage," 5–7.

12. Ibid., 8.

13. Ibid., 9–10.

14. Ibid., 12; Douglass, *Narrative*, 82.

15. "Journal of Wallace Turnage," 13–16. Douglass also portrays his fight with Covey as two hours in duration, and the turning point in his story.

16. Ibid., 19–21. Turnage's sense of distances and geography was remarkable. He declares Columbia, Mississippi, to be "twenty-seven miles" from Pickensville, Alabama, and that is exactly the mileage indicated on modern road maps. Pickensville is still a very small town, with a current population of only 169.

17. Ibid., 22–27.

18. Ibid., 28; Jacobs, *Incidents in the Life of a Slave Girl*, 148.

19. "Journal of Wallace Turnage," 29–30.

20. See Andrews, *To Tell a Free Story*, 151–66. Henry Bibb, *Narrative of the Life and Adventures of Henry Bibb* (New York: the author, 1849), 159, copy in Sterling Library, Yale University, New Haven, Conn.

21. "Journal of Wallace Turnage," 32–58.

22. Ibid., 60–66.

23. Ibid., 46–47.

24. Ibid., 51–52, 54.

25. Ibid., 48–49.

26. Ibid., 22. See Peter Cozzens, *The Darkest Days of the War: The Battles of Iuka and Corinth* (Chapel Hill: University of North Carolina Press, 1997); and Michael B. Ballard, *Civil War Mississippi, A Guide* (Jackson: University Press of Mississippi, 2000), 11–34.

27. On the Second Confiscation Act and its role in emancipation, see Allen

Guelzo, *Lincoln's Emancipation Proclamation: The End of Slavery in America* (New York: Simon & Schuster, 2004), 64–65, 112–13; and for the Preliminary Emancipation Proclamation, see Michael P. Johnson, ed., *Abraham Lincoln, Slavery, and the Civil War* (Boston: Bedford Books, 2001), 206–08.

28. R. C. Murphy to Col. Rawlins, Holly Springs, Miss., December 20, 1862; Col. Isaac C. Hawkins to Gen. J. C. Sullivan, Trenton, Miss., December 20, 1862; Gen. U. S. Grant to Col. A. Stager, Holly Springs, November 30, 1862; Gen. William T. Sherman to Brig. Gen. Alvin P. Hovey, Memphis, Tenn., October 29, 1862, in *The War of the Rebellion: A Compilation of the Official Records of the Union and Confederate Armies,* series I, vol. 17, part II (Washington, D.C.: Government Printing Office, 1887), 444, 371, 856.

29. Gen. U. S. Grant to Gen. Henry Halleck, La Grange, Miss., November 15, 1862; and Grant to Halleck, Holly Spring, Miss., January 6, 1863, in *The War of the Rebellion: A Compilation of the Official Records of the Union and Confederate Armies,* series I, vol. 17, part II, Reports (Washington, D.C.: Government Printing Office, 1886), 470, 481.

30. L. D. Sandidge, Acting Assistant Adjutant for Brig. Gen. Ruggles, "General Orders," in *The War of the Rebellion,* series I, vol. 17, part I, 638.

31. Adjutant Pleasant Smith to A.A.G. J. Thompson, January 8, 1863, Okolona, Miss., in Ira Berlin et. al., *Freedom: a Documentary History of Emancipation, 1861–1867,* series I, vol. I, *The Destruction of Slavery* (London: Cambridge University Press, 1985), 300.

32. Lawrence Ross and Lucius F. Hubbard, quoted in Cozzens, *The Darkest Days of the War,* 19.

33. "Journal of Wallace Turnage," 67–72.

34. Ibid., 73–75. See Stephen V. Ash, *When the Yankees Came: Conflict and Chaos in the Occupied South, 1861–1865* (Chapel Hill: University of North Carolina Press, 1995), 76–77, 103–07, 204–11.

35. "Journal of Wallace Turnage," 76–82.

36. Ibid., 84–86.

37. Ibid., 87–91, 94.

38. For the numbers of troops garrisoning Mobile, see Maj. Gen. Dabney Maury to James A. Seddon, Sec. of War, Mobile, Ala., February 15, 1864, *Official Records,* series I, vol. 32, part II, 739. Maury requested an additional

6,000–7,000 more troops, foodstuffs, and especially more ordnance. On the capture of Meridian as well as Forrest's victory at Okolona, see E. B. Long, *The Civil War Day by Day: An Almanac, 1861–1865* (Garden City, N.Y.: Doubleday, 1971), 464, 467; *Official Records*, series I, vol. 32, part II, 796, 800.

39. Michael W. Fitzgerald, *Urban Emancipation: Popular Politics in Reconstruction Mobile, 1860–1890* (Baton Rouge: Louisiana State University Press, 2002), 10–14; Arthur W. Bergeron, Jr., *Confederate Mobile* (Jackson: University Press of Mississippi, 1991), 3–4, 104–06. The British journalist was William Howard Russell and he visited Mobile in May 1861.

40. Bergeron, *Confederate Mobile*, 100–03; Frances Woolfolk Wallace, *Diary*, April 10, 17, 27, 1864, Southern History Collection, University of North Carolina, Chapel Hill, docsouth.unc.edu/imls.texts. Frances Wallace's husband, Philip Hugh Wallace, an officer in the Confederate army from Paducah, Kentucky, enlisted in 1861. She and a cousin engaged in a long trek across Kentucky, Tennessee, Mississippi, and down to Mobile, Alabama in 1864, trying to find and accompany their husbands at the front. Her *Diary* is an extraordinary record of that journey.

41. "Journal of Wallace Turnage," 91–93; Fitzgerald, *Urban Emancipation*, 19–20.

42. Fitzgerald, *Urban Emancipation*, 20.

43. "Journal of Wallace Turnage," 94–97.

44. Ibid., 98. On the elaborate lines of earthworks see John C. Waugh, *Last Stand at Mobile* (Abilene, Tex.: McWhiney Foundation Press, 2001), 17–18.

45. "Journal of Wallace Turnage," 99–101.

46. Numerous issues of the *Mobile Daily Tribune, Mobile Evening Telegraph, Mobile Evening News,* and *Mobile Advertiser and Register,* December 1863–August 1864, quoted in Bergeron, *Confederate Mobile,* 107–08. In April 1863, the editor of the *Advertiser and Register* chastised slaveholders for their lack of surveillance of their slaves: "It may happen some morning that somebody will wake up a nigger or two poorer than when he went to bed" (Bergeron, *Confederate Mobile,* 108).

47. "Journal of Wallace Turnage," 101.

48. Ibid., 102–03.

49. Ibid., 104; Douglass, *Narrative*, 96.

50. "Journal of Wallace Turnage," 92–93. On the battle of Mobile Bay, see Chester G. Hearn, *Mobile Bay and the Mobile Campaign: The Last Great Battles of the Civil War* (Jefferson, N.C.: McFarland, 1993), 100–20; Chester G. Hearn, *Admiral David Glasgow Farragut: The Civil War Years* (Annapolis, Md.: Naval Institute Press, 1998), 272–301; Long, *Civil War Day By Day*, 551–52; Waugh, *Last Stand at Mobile*, 23–60; James M. McPherson, *Battle Cry of Freedom: The Civil War Era* (New York: Oxford University Press, 1988), 760–61.

51. "Journal of Wallace Turnage," 105; G. Granger to Gen. E. R. S. Canby, August 5, 1864, *Official Records*, series I, vol. 39, part II, 226–27; Hearn, *Admiral David Glasgow Farragut*, 292–93.

52. Maury, quoted in Hearn, *Admiral David Glasgow Farragut*, 295; Jefferson Davis to Gen. D. H. Maury, August 9, 1864, in *Official Records*, series I, vol. 39, part II, 767; Mary Boykin Chesnut, *Diary from Dixie*, Isabella D. Martin and Myrta L. Avary, eds. (New York: Appleton, 1905), 319, 322. For the terms of the surrender and Granger's report, see *Official Records*, series I, vol. 39, part I, Reports, 417–18.

53. "Journal of Wallace Turnage," 105–06; military map of Mobile region, prepared in New Orleans, June 1, 1865, accompanying report of Gen. E. R. S. Canby, U.S. Army, University of South Alabama Archives. I am grateful to Ann Dowdle Maddox for her help in securing this map. The author toured the Foul River on a ferry boat, March 11, 2006.

54. "Journal of Wallace Turnage," 107.

55. Ball, *Fifty Years in Chains*, 367–68; *Mark* 1: 9–13.

56. "Journal of Wallace Turnage," 108.

57. Ibid., 109–10, 106.

58. Ibid., 111–12.

59. Ibid., 113–14.

60. Ibid., 115–18. The black folk lyrics "trouble in mind," or sometimes "troubled in mind," were first perhaps recorded in the 1920s and 1930s as a blues song. But their roots are much older in the slave spiritual, "I am a-trouble in de mind,/ Oh I am a-trouble in de mind;/ I ask my Lord what shall I do,/ I am a-trouble in de mind." See Lawrence W. Levine, *Black Culture and Black Consciousness: Afro-American Folk Thought from Slavery to Free-*

dom (New York: Oxford University Press, 1977), 230. Leon Litwack takes the lyric as the title of Litwack, *Trouble in Mind: Black Southerners in the Age of Jim Crow* (New York: Knopf, 1998), vii, 326. The second version of the song, which is the actual phrase Turnage used, is from Federal Writers' Project, Works Progress Administration, Folklore Subjects, folder 31 (songs), Arkansas State Archives, Little Rock, Ark. (Litwack, *Trouble in Mind*, n. 1, 541). Whether Turnage had ever heard such a lyric before writing the phrase in his narrative is impossible to know; likely he had, but he is also simply one of thousands who lived the experience and used the words that eventually made their way into a blues song: "Oh, Lord, I'm troubled in mind!/ I want you to ease my troubled mind."

61. "Journal of Wallace Turnage," 119–20. I speculate that the actual day of Turnage's liberation may have been August 24, 1864, because the fall of Fort Morgan occurred on the afternoon of August 23, three miles across the bay from Fort Gaines. General Granger turned over the command of Morgan to General George Gordon and, according to an itinerary in the *Official Records*, "proceeded to New Orleans," stopping in all likelihood on Dauphin Island at Fort Gaines the following day before his departure. Moreover, another itinerary indicates that Union forces occupied Cedar Point on August 25, the very site from which Turnage had launched his rowboat. See *Official Records*, series I, vol. 39, part I, Reports, 422. This also conforms with Turnage's memory that his escape occurred in the "latter" part of August.

62. "Journal of Wallace Turnage," 121. Turnage erroneously identifies Junius Turner as "Julius." On Granger, see Waugh, *Last Stand at Mobile*, 62–63. For his orders about black regiments, G. Granger, August 20, 1864, *Official Records*, series I, vol. 39, part I, Reports, 418–19. Granger's order came in the urgency of the siege of Fort Morgan.

63. Junius Thomas Turner, pension application files, certificates number 225515 and 184214, National Archives Record Administration (NARA), Washington, D.C. Turner lived for some time on Whidbey Island, Washington Territory, and he may have volunteered in the Indian Wars, 1855–56. See *Washington Post*, September 5, 1921; and Rep. Albert Johnson to Commissioner of Pensions, December 23, 1926, Laura A. Turner, Widow's Pension Application File number 958886, NARA. Also see Junius Thomas Turner, 3rd

Maryland Cavalry, Military Service Record, NARA. Among the military positions in which Turner served were Acting Assistant Adjutant General in the Military District of Western Mississippi; Assistant Inspector General of Cavalry, Department of the Gulf; a company commander at Fort Adams, Mississippi; and Judge Advocate in Natchez, Mississippi. He experienced malarial symptoms and other maladies that forced him to return East before his regiment departed.

64. Turner Pension Files, numbers 225515 and 184214.

65. "Journal of Wallace Turnage," 120.

Chapter Three: Unusual Evidence

1. W. E. B. Du Bois, *Black Reconstruction in America, 1860–1880* (1935 rpt. New York: Atheneum, 1962), 80.

2. Ibid., 238.

3. W. E. B. Du Bois, *The Black North in 1901: A Social Study* (1901 rpt. New York: Arno Press, 1969), 2.

4. Hennessey, *Return to Bull Run*, 441, 450–51; *Official Records*, series I, vol. 12, part III, 787, 797.

5. Ernest B. Furguson, *Freedom Rising: Washington in the Civil War* (New York: Knopf, 2004), 197–98, 256–57; Constance McLaughlin Green, *The Secret City: A History of Race Relations in the Nation's Capital* (Princeton, N.J.: Princeton University Press, 1967), 62–64; James Borchert, *Alley Life in Washington: Family, Community, Religion, and Folk Life in the City, 1850–1970* (Urbana: University of Illinois Press, 1980), 26–28; Long, *The Civil War Day by Day*, 258.

6. See Elizabeth Clark-Lewis, ed., *First Freed: Washington, D.C. in the Emancipation Era* (Washington, D.C.: Howard University Press, 2002). "History of Shiloh," www.shilohbaptist.org/ded. On the building of Shiloh Baptist, see *Washington Bee*, January 12, 1884, June 8, 1889. On fairs and excursions, see *Washington Bee*, October 25, November 8, and December 6, 1884, and August 22, 1885.

7. John W. Cromwell, "The First Negro Churches in the District of Columbia," *Journal of Negro History*, 1922, vol. 7, no. 1, 89; *Washington Post*, August 4, September 15, 1879, October 3, 1900; *Washington Bee*, August 1 and January 22, 1887.

8. The split in the Shiloh congregation seemed to develop in the final years and at the death of Rev. William Walker. For Walker's extraordinary memorial service, at which various ministers competing for his pulpit spoke, see *Washington Bee*, January 25, 1890. John and Annie would surely have attended such an event. One press report claimed that "every woman in the church" favored a Rev. W. H. Scott, while all the men opposed him. See *Washington Bee*, November 8, 1890. And as late as 1891, another article reported that "disorder and confusion" still reigned at Shiloh and that many good Baptists were "disgusted" with the split in the church. See *Washington Bee*, May 9, 1991. The role the Washingtons may have played in this dispute cannot be determined.

9. *Boyd's Directory of Washington and Georgetown*, William H. Boyd, comp. (Washington, D.C.: Hudson Taylor Bookstore, 1864), 272; Clark-Lewis, ed., *First Freed*, 79–83.

10. Clark-Lewis, ed., *First Freed*, 92–95.

11. Ibid., 86–92; *Washington Evening Star*, April 14, 1866.

12. Craig A. Schiffert, "Stepping Toward Freedom: An Historical Analysis of the District of Columbia Emancipation Day Parades, 1866–1900," in Clark-Lewis, ed., *First Freed*, 111–31; Frederick Douglass, "Address Delivered on the Twenty-sixth Anniversary of Abolition in the District of Columbia," April 16, 1888, Douglass Papers, LC, reel 16.

13. Frederick Douglass, "Thoughts and Recollections of the Antislavery Conflict," speech, undated, Douglass Papers, LC, reel 16.

14. On the Fourteenth Amendment as "second founding," see Garrett Epps, *Democracy Reborn: The Fourteenth Amendment and the Fight for Equal Rights in Post-Civil War America* (New York: Henry Holt, 2006).

15. She is not in the 1870 census; moreover, no death certificate for her survives and the District of Columbia did not begin recording such documents until 1874.

16. "The Death of Our Little Johnnie, Who Died March 2nd, 1865, at 1:20 pm, aged 8 months and 25 days," May 10, 1865, Washington Papers (MHS).

17. "The Death of Our Little Johnnie"; Walt Whitman, "When Lilacs Last in the Dooryard Bloom'd," *Leaves of Grass* (Garden City, N.Y.: Doubleday, 1926), 275, 281.

18. *Boyd's Directory of Washington, Georgetown, and Alexandria*, 497.

19. The Freedmen's Bank account, deposit of $150.00, September 18, 1867, www.content.gale.ancestry.com. Washington actually opened the bank account on behalf of the Daughters of Shiloh Church. Also, in 1870, Washington is listed as Secretary for the Children's Band of Hope, either a church or civic aid society. See Freedmen's Bank account, deposit of $80.00, July 5, 1870, www.content.gale.ancestry.com. Certificate of Death, John M. Washington, Commonwealth of Massachusetts, Cohasset, Norfolk County, issued March 12, 1918. *Boyd's Directory of Washington, Georgetown, and Alexandria,* 1867, 567; 1868, 446; 1869, 465; 1870, 383.

20. Certificate of Death, # 24133, May 3, 1880, Sarah Tucker, Health Department, District of Columbia. On the birthdates of all the Washington children, see "Family Record," Washington Papers (MHS). U.S. Census, 1870, 235–36. In the 1870 and 1880 censuses, John and Annie are listed as "m" for mulatto, but in the 1900 census as "b" for black. By 1880, Annie is listed as "keeping house." U.S. Census, 1880, 36; U.S. Census, 1900, 17A.

21. U.S. Census, 1890; 1910.

22. *Chicago Defender,* October 1, 1910, July 29, 1911, January 27, 1912, February 17, 1912, June 8, 1912, July 20, 1912, August 27, 1912, November 23, 1912, April 13, 1913, March 28, 1914, September 15, 1917, March 11, 1922, February 5, 1927. Mabel Washington obituary, *Chicago Defender,* May 27, 1933. William and Louisa had a son, William Earl Washington, who as a youth was a high school basketball star in the 1920s at Hyde Park High School. Earl became a state employee, working for the Illinois Labor Department. He died in an auto accident in 1972; his only descendents were two stepchildren. See *Chicago Defender,* November 2 and 4, 1972.

23. Ralph Ellison, "Going to the Territory," in *Going to the Territory* (New York: Vintage, 1987), 131.

24. *Who's Who of the Colored Race* (Chicago: 1915), 277; obituary, Benjamin Washington, *Washington Post and Times Herald,* December 10, 1957. On M Street High School and the black school district of Washington, see Jacqueline M. Moore, *Leading the Race: The Transformation of the Black Elite in the Nation's Capital, 1880–1920* (Charlottesville: University Press of Virginia, 1999), 86–93.

25. Rayford W. Logan, *Howard University: The First Hundred Years, 1867–1967* (New York: NYU Press, 1969), 133; Moore, *Leading the Race*, 87; *Washington Post*, December 10, 1957; *The Black Washingtonians: The Anacostia Museum Illustrated Chronology* (Hoboken, N.J.: John Wiley & Sons, 2005), 113. "First Public Appearance of the Washington High School Cadets," March 4, 1893, list of officers and participants, prepared by Benjamin Washington, in Washington Papers (MHS).

26. *Chicago Defender*, January 15, 1927; *Washington Post and Times Herald*, April 1 and 3, 1955. At his retirement from teaching, a banquet was held in Benjamin's honor, attended by many graduates of M Street High School from the classes of 1893–96. This was surely a measure of Benjamin's distinction in the community. See *Washington Post*, November 5, 1943.

27. U.S. Census, District of Columbia, 1910, p. 2470; *New York Times*, June 14, 2004.

28. On Nineteenth Street Baptist, see commemorative pamphlet, *One Hundredth Anniversary of the Nineteenth Street Baptist Church, Washington, D.C., 1839–1939*; and Mark Tucker, *Ellington: The Early Years* (Urbana: University of Illinois Press, 1991), 22.

29. *Washington Post*, December 10, 1957.

30. David H. Wadsworth, "Information Sought about Washington Family," *Cohasset Mirror*, April 14, 1982; Assessor's Report, form number COH 1233, 321 North Main Street, Cohasset, for Massachusetts Historical Commission, April 1996, copy in Cohasset Historical Society; Assessor's Map number 14, Cohasset Historical Society.

31. For Thomas Loney, see photograph, in David Wadsworth, Paula Morse, and Lynne DeGiacomo, *Images of America: Cohasset* (Portsmouth, N.H.: Arcadia, 2004), 107. Loney was born a slave in Millenbeck, Virginia, in 1833. Assessor's Map number 14, Cohasset Historical Society. The Bowsers lived at 302 North Main Street.

32. On the process of sectional reconciliation and its impact on Civil War memory, as well as the character of African American memory of emancipation and the war, see David W. Blight, *Race and Reunion: The Civil War in American Memory* (Cambridge, Mass.: Harvard University Press, 2001).

33. Certificate of Death, Commonwealth of Massachusetts, John M. Washington, filed March 18, 1918; Form E-Burial Ground, Inventory Form, number 802, for Woodside Cemetery, Massachusetts Historical Commission, March 2006, copy in Cohasset Historical Society; Robert Lowell, "For the Union Dead," in *Norton Anthology of American Literature* (New York: Norton, 1980), vol. I, 842.

34. DuBois, *Black Reconstruction*, 238; "Journal of Wallace Turnage," 122. Turnage's mother is listed in 1870 as Courtney Tyson. U.S. Census, Pitt County, N.C., 1870, 42.

35. U.S. Census, New York City, 1870, 21; James Weldon Johnson, *Black Manhattan* (rpr. 1930; New York: Atheneum, 1968), 58–59; and Gerald W. McFarland, *Inside Greenwich Village: A New York City Neighborhood, 1898–1918* (Amherst: University of Massachusetts Press, 2001), 11–13.

36. Freedmen's Bank accounts, number 2090, January 17, 1871; number 4017, September 12, 1871; and number 5184, November 12, 1872, www.content.gale.ancestry.com.

37. Paul Lawrence Dunbar, in Shelley Fisher Fishkin and David Bradley, eds., *The Sport of the Gods and Other Essential Writings* (1902 rpr. New York: Modern Library, 2005), 356–57.

38. Freedmen's Bank account, November 21, 1872, number 5184, Wallace Turnage; Freedmen's Bank account, March 20, 1874, number 6801, Nelson Hart, www.content.gale.ancestry.com; Certificate of Marriage, State of New York, Wallace Turnage and Sarah Bird, number 2794, May 10, 1875, Bureau of Records and Vital Statistics, Health Department of the City of New York; U.S. Census, 1850, R522, 273.

39. U.S. Census, New York City, 1870, 21; U.S. Census, 1880, New York City, 27. Jacob Riis, *How the Other Half Lives* (1890 rpr; Boston: Bedford Books, 1996), 157, 161–62; Stephen Crane, "Stephen Crane in Minetta Lane," in Fredson Bowers, ed., *Stephen Crane: Tales, Sketches, and Reports* (Charlottesville: University Press of Virginia, 1973), 400–04; and McFarland, *Inside Greenwich Village*, 11–18.

40. U.S. Census, Jersey City, N.J., 1880, 47; Birth Certificate, number 177295, Ida Turnage, February 13, 1876, Bureau of Vital Statistics, Health Department, New York.

41. David Quigley, *Second Founding: New York City, Reconstruction and the Making of American Democracy* (New York: Hill and Wang, 2004), 91–100, 104–08.

42. Du Bois, *The Black North in 1901*, 16.

43. *New York Freeman,* September 18, 1886, October 16, 1886, and February 26, 1887; *New York Age,* March 10, 1888, March 2, 1889, February 21, 1891.

44. On the New York City Memorial Day commemorations of 1877 and 1878, see Blight, *Race and Reunion,* 87–93; *New York Tribune,* May 29–31, 1877. Douglass's speech is "Speech in Madison Square in Honor of Decoration Day," May 30, 1878, Douglass Papers, Library of Congress, reel 15.

45. Death Certificate, number T19, Abbie Turnage, September 16, 1884; Death Certificate, number T18, Wallace Turnage, Jr., September 19, 1884, Bureau of Records and Vital Statistics, Health Department, City of New York.

46. Samuel H. Preston and Michael R. Haines, *Fatal Years: Child Mortality in Late Nineteenth Century America* (Princeton, N.J.: Princeton University Press, 1991), 90–96, 141–46, 208–10. Childhood mortality was considerably lower for Southern rural blacks than for Northern urban dwellers like the Turnages. Preston and Haines calculated the overall national rate of childhood mortality for blacks to be 58 percent higher than that of whites. Also see Reynolds Farley, *Growth of the Black Population: A Study of Demographic Trends* (Chicago: Markham Publishing, 1970), 61–62.

47. "Narrative of Wallace Turnage," 123; Acts 8: 23; Matthew 27: 34. The passage in Acts reads: "For I perceive that thou art in the gall of bitterness, and in the bond of iniquity."

48. Death Certificate, State of New York, number 28553, Courtney Hart, October 5, 1898, Bureau of Records and Vital Statistics, Health Department, City of New York.

49. Certificate of Marriage, Wallace Turnage and Sarah Bohannah, number 13096, State of New York, November 19, 1889, Sanitary Bureau, Division of Vital Statistics, Health Department, City of New York.

50. Ralph Ellison, "The World and the Jug," in Ellison, *Shadow and Act* (New York: Vintage, 1964), 115.

51. Turnage obituary, *Jersey Journal,* October 6, 1916, 13; Death Certificate, number 8060, Wallace Turnage, State of New Jersey, Bureau of Vital Statistics;

Declarations for Widow's Pension, Sarah Turnage, March 20, 1919, April 8, 1919; letter, Thomas P. Kenny, agent of Philadelphia Underwriters, Department of Insurance Company of North America, to Bureau of Pensions, on behalf of Sarah Turnage, March 20, 1919; Pension application, Sarah Turnage, May 31, 1919; reply letters from Pension Commissioner to Sarah Turnge, April 21, 1919, June 20, 1919, National Archives, number 1138617, can number 3472, bundle number 50. Sarah died of "old age" at sixty-eight in 1921. Her obituary indicated that "no members of her immediate family survive," implying that Turnage's daughters, Sarah and Lydia, did not develop close ties with their stepmother. Obituary, "Mrs. Sarah Turnage," *Jersey Journal,* January 21, 1921, 9. For the one major effort that did exist to bring about slave pensions, however unsuccessful and led by a black woman, Callie House, see Mary Frances Berry, *My Face Is Black Is True: Callie House and the Struggle for Ex-Slave Reparations* (New York: Knopf, 2005).

52. U.S. Census, 1900, Hudson County, N.J.; U.S. Census, 1910, Hudson County, N.J., sheet number 3551; U.S. Census, 1910, Brooklyn, N.Y., sheet number 6B; Registration Card and Report, William Turnage, 1918, Local Board number 81, Public School No. 64, Brooklyn, N.Y.; Death Certificate, number 965, William Turnage, June 1, 1928, Bureau of Records, Department of Health, City of New York.

53. U.S. Census, 1910, sheet number 3551; Death Certificate, number 444, Thomas Daniel Connolly, November 16, 1964, Connecticut State Department of Health; Death Certificate, number 446, Lydia Connolly, October 30, 1884, State of Connecticut Department of Health Services.

54. Box and postcard in Turnage Papers, Historical Society of the Town of Greenwich, box 1.

Chapter Four: The Logic and the Trump of Jubiliee

1. Frederick Douglass, *Life and Times of Frederick Douglass* (rpr. 1882; New York: Collier, 1962), 352–53; David W. Blight, *Frederick Douglass' Civil War: Keeping Faith in Jubilee* (Baton Rouge: Louisiana State University Press, 1989), 106.

2. Douglass, *Life and Times,* 353–54.

3. Benjamin Quarles, *The Negro in the Civil War* (Boston: Little Brown, 1953), 174–76. John Greenleaf Whittier wrote a poem, often converted into song, and rooted in the black music he heard and witnessed in the sea islands of low-country Georgia, entitled "Song of the Negro Boatman." See W. J. Sidis, comp., "America's Search for Liberty in Song and Poem," 1935, www.sidis .net/asl.htm.

4. Pleasant Smith to A.A.G.J. Thompson, January 8, 1865, Okolona, Mississippi, headquarters of the Confederate Department of Mississippi and East Louisiana, in Ira Berlin et al., eds., *Freedom: A Documentary History of Emancipation, 1861–67*, series 1, vol. 1 (London: Cambridge University Press, 1985), 300; Cam Walker, "Corinth: The Story of a Contraband Camp," *Civil War History*, vol. 20 (March 1974), 5–22.

5. Douglass, *Life and Times*, 353.

6. This mixture of factors relies heavily on the work of Ira Berlin and others. For the argument, see Ira Berlin, Barbara J. Fields, Steven F. Miller, Joseph P. Reidy, and Leslie S. Rowland, *Slaves No More: Three Essays on Emancipation and the Civil War* (London: Cambridge University Press, 1992), 5–6 and throughout. Emancipation, these authors argue, was "a varying, uneven, and frequently tenuous process." But they do conclude that slaves themselves must be viewed in the end as the "prime movers" in the story. For a somewhat contrary view to this, which gives Lincoln and the Union armies more weight in the balance of responsibility for how the slaves attained freedom, see James M. McPherson, "Who Freed the Slaves?" *Proceedings of the American Philosophical Society*, vol. 139 (March, 1995), 1–10. For a fuller, assertive claim about Lincoln's centrality in emancipation and his consistent aim to free slaves, see Allen Guelzo, *Lincoln's Emancipaton Proclamation: The End of Slavery in America* (New York: Simon & Schuster, 2004); and Richard Striner, *Father Abraham: Lincoln's Relentless Struggle to End Slavery* (New York: Oxford University Press, 2006).

7. *Official Records*, series 2, vol. 1, 750.

8. Benjamin F. Butler to General-in-Chief of the Army, Fortress Monroe, Va., May 27, 1861, in Berlin et al., eds., *Destruction of Slavery*, 72–73; *Official Records*, series 1, vol. 2, 648–52.

9. Berlin et al., *Slaves No More*, 21.

10. Ibid., 22–23, 27–28; David W. Blight, ed., *When This Cruel War Is Over: The Civil War Letters of Charles Harvey Brewster* (Amherst: University of Massachusetts Press, 1992), 18–19. Brewster was a lieutenant in the Tenth Massachusetts, stationed in Camp Brightwood, outside Washington, D.C., in the winter of 1862. "He was the only slave his master had," wrote Brewster of his contraband, "and his master will never have him again if I can help it." The exclusion policy caused a three-month-long, bitter dispute in Brewster's regiment and nearly "a state of mutiny." See letters of March 5, 8, and 12, 1862, 93–97.

11. Berlin et al., *Slaves No More*, 30–31; and Michael J. Kurtz, "Emancipation in the Federal City," *Civil War History*, vol. 24, no. 3 (September, 1978), 250–67. The compensated emancipation process in the District was administered by a three-man board of commissioners with a budget of $1 million. Great anxiety ensued initially in the city, but in the end, the process worked relatively well. The commissioners received 966 petitions for compensation for 3,100 slaves; they accepted 909 in their entirety, 21 in part, and rejected 36. See Kurtz, "Emancipation in the Federal City," 264. Washington *Evening Star*, March 31, 1862, and *National Intelligencer*, April 4, 1862, both quoted in David Taft Terry, "A Brief Moment in the Sun: The Aftermath of Emancipation in Washington, D.C., 1862–1869," in Clark-Lewis, *First Freed*, 76–77.

12. Frederick Douglass to Montgomery Blair, September 16, 1862, in Phillip Foner, ed., *Life and Writings of Frederick Douglass* (New York: Atheneum, 1950), vol. 3, 288–89.

13. Frederick Douglass, "The War and How to End It," speech at Corinthian Hall, Rochester, N.Y., March 25, 1862, in *Douglass Monthly*, April 1862.

14. Lincoln, in Basler, ed., *Collected Works of Abraham Lincoln*, vol. 5, 317–19. Shortly after the meeting, twenty of twenty-eight border-state Congressmen voted to reject Lincoln's offer of gradual, compensated emancipation.

15. *United States Statutes At Large* (Washington, D.C.: Government Printing Office, 1863), vol. 12, 589–92; Berlin et al., *Slaves No More*, 40–41.

16. Howard K. Beale, ed., *Diary of Gideon Welles, Secretary of the Navy Under Lincoln and Johnson* (New York: Norton, 1960), vol. 1, 70.

17. Du Bois, "Abraham Lincoln," *Crisis*, September 1922, *in W. E. B. Du Bois: Writings* (New York: Library of America, 1986), 1196, 1198. On Douglass and

colonization, see Blight, *Frederick Douglass' Civil War,* 122–47. For Lincoln's early colonizationist thought, see Basler, ed., *Collected Works of Lincoln,* vol. 2, 132, 255, 298–99, 409–10.

18. Lincoln to Greeley, August 22, 1862, in Johnson, ed., *Abraham Lincoln,* 204–05. On the Greeley letter, see Guelzo, *Lincoln's Emancipation Proclamation,* 132–37. Guelzo calls Lincoln's response to Greeley's challenge a measure of his "passionless, logical progress" on emancipation.

19. "Preliminary Emancipation Proclamation," September 22, 1862, in Johnson, ed., *Abraham Lincoln,* 206–08.

20. "Annual Message to Congress," December 1, 1862, in Johnson, ed., *Abraham Lincoln,* 209–17.

21. Lincoln, quoted in Francis B. Carpenter, *Six Months at the White House with Abraham Lincoln: The Story of a Picture* (New York: Hurd and Houghton, 1866), 90; "Emancipation Proclamation," in Johnson, ed., *Abraham Lincoln,* 218–19.

22. See Bruce Levine, "Fight and Flight: The Wartime Destruction of Slavery," in David W. Blight, ed., *Passages to Freedom: The Underground Railroad in History and Memory* (Washington, D.C.: Smithsonian Books, 2004), 211–30; Ira Berlin et al., *The Documentary History of Emancipation, 1861–1867: Black Military Experience,* series II (Cambridge: Cambridge University Press, 1982), 1–34.

23. Peter Cooper, quoted in James M. McPherson, *The Negro's Civil War: How American Negroes Felt and Acted During the War for the Union* (New York: Vintage, 1965), p. 65; freedman quoted in Leon F. Litwack, *Been in the Storm So Long: The Aftermath of Slavery* (New York: Knopf, 1979), 21; "W. J. G.," in *Portage County Democrat,* Portage, Ohio, October 22, 1862, in Victor B. Howard, "The Civil War in Kentucky: The Slave Claims His Freedom," *Journal of Negro History,* vol. 67, no. 3 (Autumn 1982), 247.

24. Charles C. Jones to Eliza G. Robarts, July 5, 1862, and Jones to Charles C. Jones, Jr., July 10, 1862, in Robert M. Myers, ed., *The Children of Pride: A True Story of Georgia and the Civil War* (New Haven, Conn.: Yale University Press, 1972), 925–26, 929–30.

25. Edmonston and Clanton quoted in James L. Roark, *Masters Without Slaves: Southern Planters in the Civil War and Reconstruction* (New York:

Norton, 1977), 82, 89; John F. Andrews to Mrs. Clement Claiborne Clay, July 10, 1863, quoted in David Williams, "The 'Faithful Slave' Is About Played Out': Civil War Slave Resistance in the Lower Chattahoochee Valley," *Alabama Review*, no. 52 (April 1999), 86; Mary Jones, journal, January 21, 1865, in Myers, ed., *Children of Pride*, 1247.

26. William Faulkner, *The Unvanquished* (rpr. 1934; New York: Vintage, 1966), 94–95.

27. Deposition of Octave Johnson, American Freedmen's Inquiry Commission, February 1864, in Berlin et al., *Destruction of Slavery*, 217.

28. Quoted in Litwack, *Been in the Storm So Long*, 212–13; Toni Morrison, *Beloved* (New York: Knopf, 1989), 273.

29. Ira Berlin, Barbara J. Fields, Steven F. Miller, Joseph P. Reidy, and Leslie S. Rowland, eds., *Free At Last: A Documentary History of Slavery, Freedom, and the Civil War* (New York: New Press, 1992), 185–86; Berlin et al., *Slaves No More*, 117–20, 131–33, 158–61.

30. Berlin et al., eds., *Free At Last*, 186–200.

31. Walker, "Corinth: The Story of a Contraband Camp," 5–15; Rome, *Georgia Weekly Courier*, September 19, 1862, in Paul D. Escott, "The Context of Freedom: Georgia's Slaves During the Civil War," *Georgia Historical Quarterly*, vol. 58 (Spring 1974), 84.

32. Walker, "Corinth: The Story of a Contraband Camp," 15–22.

33. Ibid., 20–22.

34. Eaton, quoted in John Cimprich, *Slavery's End in Tennessee, 1861–1865* (Knoxville: University of Tennessee Press, 1985), 49.

35. Harriet Jacobs, "Life Among the Contrabands," *Liberator*, September 5, 1862, quoted in Jean Fagan Yellin, *Harriet Jacobs: A Life* (New York: Basic Books, 2004), 159.

36. Harriet Jacobs to John Sella Martin, Alexandria, Va., April 13, 1863, quoted in Yellin, *Harriet Jacobs*, 167.

37. Nathaniel Hawthorne, "Chiefly About War Matters," *Atlantic Monthly* (July 1862), in Yellin, *Harriet Jacobs*, 163; Edward Dicey, *Spectator of America*, Herbert Mitgang, ed. (Chicago: Quadrangle, 1971), 154.

38. Jacobs, "Life Among the Contrabands"; and Jacobs to John Sella Martin, April 13, 1863, all quoted in Yellin, *Harriet Jacobs*, 163–64.

39. Rev. John Jones to Mrs. Mary Jones, August 21, 1865, in Myers, ed., *The Children of Pride*, 292.

40. On "refugeeing," see Clarence L. Mohr, *On the Threshold of Freedom: Masters and Slaves in Civil War Georgia* (Athens: University of Georgia Press, 1986), 99–119.

41. Ibid., 103–19.

42. See Lynda J. Morgan, *Emancipation in Virginia's Tobacco Belt, 1850–1870* (Athens: University of Georgia Press, 1992), 110.

43. Ibid., 114, 118.

44. Berlin et al., *Slaves No More*, 178–79; *Continental Monthly*, January and February 1862, in V. Jacque Voegeli, "A Rejected Alternative: Union Policy and the Relocation of Southern 'Contrabands' at the Dawn of Emancipation," *Journal of Southern History*, vol. 69 (November 2003), 765.

45. Berlin et al., *Slaves No More*, 179–80.

46. Emily Waters to "my Dear Husband," July 16, 1865, in Berlin et al., eds., *Free At Last*, 525–26.

47. John Boston to Elizabeth Boston, Upton Hill, Va., January 12, 1862, in ibid., 29–31. It is not likely that Elizabeth ever read John's letter; it was intercepted and ended up in the hands of a committee of the Maryland House of Delegates, which gave it to Gen. George B. McClellan, demanding that Boston be returned to his "loyal" owner.

48. Richard Rodriguez, "An American Writer," in Werner Sollors, ed., *The Invention of Ethnicity* (New York: Oxford University Press, 1989), 7–8.

John Washington, "Memorys of the Past"

1. Quilts or bed coverings.

2. Deuteronomy 34:1–4.

3. He likely means "etc."

4. From this expression we can discern that Washington wrote his narrative in 1873. Nowhere else does he date the work.

5. Commonly known as "The Children's Prayer."

6. Workers who cut and bound wheat and other grains. This farm likely also grew tobacco.

7. Large barrels; a hogshead is an official agricultural liquid measure of 63 gallons.

8. A children's grammar book.

9. A slave coffle in the domestic slave trade from Virginia to states in the deep South and Southwest.

10. John's four siblings, probably all born on the Brown farm in Orange County, and with a different father than John.

11. This three-part racial segregation of seating in the church reflects the three-tiered racial system of the slave states of the antebellum South. On the streets outside in Fredericksburg, slaves and free blacks mingled with regularity.

12. Rev. Richard H. Phillips, an Episcopal minister and principal, since 1848, of the Virginia Female Institute (now Stuart Hall), a girls' boarding school. He was married to Catherine Taliaferro's niece, Eleanor Thom.

13. This section on the parting with his mother and his conversion of that experience into the moment that most kindled his "hatred" of slavery and desire to escape reads remarkably like the antebellum slave narratives that exploited just such scenes for effective antislavery propaganda.

14. One of America's early great illustrated magazines, published in New York, 1850–1889. The magazine was just in its first few years when John remembers eagerly reading it.

15. Sarah had several siblings, including George, Henry, Jenny, and Laura. George was born approximately 1821.

16. This may be *Cowley's System of Practical Penmanship* (Pittsburgh: A. H. English & Co., c.1862). The author is Alexander Cowley. An older edition likely existed.

17. William Walker, minister in Fredericksburg, who will eventually move north to Washington as part of the wave of freedmen and found the Shiloh Baptist Church in the District of Columbia.

18. Washington's eloquent remembrance of his struggle to learn to read and especially to write is strikingly similar to that of Frederick Douglass in *Narrative*, 65–71.

19. The Rappahannock River was a source and an avenue of both temporary and ultimate freedom for Washington.

20. Poison oak disease.

21. Washington must have observed many blacks in the first two of these institutions. Here we can see some influence of abolitionist writings as he exploits opportunities to convert his memory and experience into an antislavery statement.

22. This section is a remarkable example of Washington's ability to mix his sense of humor four paragraphs above (considering whether he would commit the "disgraceful act" of escape) with rich abolitionist polemic (the tearful partings with mother and siblings) and the detailed memory of the trains and accident resulting in the death of the black worker. It represents his ambition as a writer despite his spelling. The "Niggers Car" was a Jim Crow car before the latter name would have been applied. Washington's sense of irony and symbolism, so often on display in the narrative, is powerful here.

23. These two paragraphs are a prime example in the narrative of Washington's use of nature as a metaphor for freedom, which had been a common theme in many slave narratives. See Frederick Douglass's famous metaphor of the sailing ships on Chesapeake Bay, in *Narrative*, 83–84.

24. In the Christian calendar, the week beginning with the seventh Sunday after Easter, commemorating the descent of the Holy Spirit upon the disciples of Christ.

25. John was fifteen years old and Annie, eleven.

26. From a hymn by C. M. Azmon, in a collection, "Southern Harmony," by William Walker, 1835, www.ccel.org/ccel/walker,harmony.html. As a poem the passage was also used in many sermons at least as early as the 1840s, and its origins may be in romantic poetry. Washington would have learned it as a common expression of faith either in church or from devotional readings.

27. Both Harriet Jacobs and Frederick Douglass also employed the metaphor of prison, cell, or tomb in their autobiographies. See Jacobs, *Incidents in the Life of a Slave Girl*; Douglass, *Narrative*, 89.

28. He means 1859.

29. Misspelled, William T. Hart.

30. The task system of labor management was used widely from cotton plantations to urban factories. Sometimes it allowed slaves time for their own garden production, as on plantations, or money, as in Washington's "hired out" case.

31. Washington's depiction of his "hiring out" experiences is one of the most revealing in any slave narrative, and this passage is indicative of how widely it was practiced in towns and cities. The tavern owners were a Greek immigrant, Speredone Zetelle, and a German immigrant, Caspar Wendlinger.

32. George Peyton and James Mazeen; both joined the 30th Virginia Infantry that winter or spring.

33. Salisbury, North Carolina.

34. Rather than the term "slave," Washington employs the more lofty label, "man servant," as if to give himself and his comrades a heightened ranking in this moment of imminent liberation.

35. Rufus King, former engineer, newspaper editor, and attorney general of New York; Union general, commander of the First Division, Pope's army, defeated at Groveton in Second Manassas campaign, August 28, 1862. Irvin McDowell, Union general, commander of the Union army at First Bull Run, July, 1861, and commander of the Third Corps of Pope's army, Second Manassas campaign. Christopher C. Augur, Union general, brigade commander, Army of Virginia, March–July, 1862, division commander wounded severely at the battle of Cedar Mountain, August 8, 1862. Washington's memory for the details of names, dates, and places is remarkably accurate.

36. Washington's sense of detail and drama in covering this scene demonstrates a remarkable example of his ambition as a writer and reflects his longterm residence in Fredericksburg. It also depicts a stunning example of how the nature of the war and the impending emancipation were realities no one could any longer hide from the slaves themselves.

37. The passage in Congress on April 16 of emancipation in the District of Columbia.

38. Washington may be misremembering Ladd's name. Ladd does not appear in the 21st New York's regimental history.

39. Eliza Butler, a free black laundress in Falmouth with eight children. Washington is surely aware by the time he wrote this that such a scene of white Union soldiers and a few freshly escaped freedmen bedding down together in a free black woman's humble house was a representation of the revolutionary character the war had taken on in 1862.

40. A hymn by Daniel Read, published as early as Read, *The American Singing Book* (five editions, 1785–96), and most recently in *Daniel Read: Collected Works*, Karl Kroeger, ed. (Madison, Wisc.: American Musicological Society, 1995).

41. An eighteenth-century hymn, written by Ignaz Pleyal, a student of Joseph Haydn, 1791. Known as the "Masonic Dirge," the popular melody with varying lyrics appears in many Christian hymnals. Washington again shows his wide familiarity with church music and devotions. See www. Masonmusic .org/pleyallost.html.

42. These bridges built by the army were known as pontoon bridges. Sometimes spelled "ponton" in modern military histories. See Mark M. Boatner, ed., *The Civil War Dictionary*, rev. ed. (1959 rpr. New York: David McKay Company, 1988), 658.

43. Reuben T. Thom, the brother-in-law of John's owner, Catherine Ware Taliaferro. Thom was the former owner of John's mother and grandmother. Washington's role as guide on horseback was rare for a former slave in the Union army. It is indicative of his skills and value to the Union officers among whom he had landed.

44. Battle of Front Royal, fought May 30, 1862, as part of Stonewall Jackson's Shenandoah Valley campaign. Washington may have read some campaign histories after the war while preparing to write his narrative.

45. John Singleton Mosby (in legend known as the "Gray Ghost"), a Virginian, led a band of irregular Confederate cavalry during the war and became famous for daring raids on Union encampments and even executions of Union officers.

46. Fauquier White Sulphur Springs, considered medicinal waters as early as the eighteenth century, near Warrenton, Virginia.

47. "Contraband" was the term applied as early as August 1861 for escaped slaves allowed into Union lines as confiscated "property" of the enemy. The term was eventually in wide use as a label for any freed slaves.

48. The battle of Cedar Mountain, part of the Second Manassas campaign, was fought on August 9, 1862, with nearly 3,700 casualties on the two sides. Washington's dates here are therefore accurate. Nathaniel P. Banks was a former

congressman and governor of Massachusetts, corps commander in the Army of the Potomac, defeated by Confederate Stonewall Jackson at Cedar Mountain. John Pope, former military engineer and explorer in the West, was commander of the Army of Virginia, instituted by Lincoln to defend northern Virginia in summer 1862. Pope was disastrously defeated at Second Manassass, August 29– September 1.

49. Here, Washington's writing resembles a widespread genre of postwar veterans' narratives that depict the adventures of wartime service. See Blight, *Race and Reunion,* 140–210.

50. "Servants" meant largely black former slaves.

51. Washington must have heard this via the slave and former slave grapevine. I cannot independently verify the offer of the $300 reward, but John was extremely valuable to Mrs. Taliaferro. An even greater fear was likely retribution against his pregnant wife, Annie.

52. Here, Washington writes like a romantic war correspondent covering the campaign at the front. By 1873, when he wrote the narrative, it is possible he had read some of the many war histories published in the immediate wake of the conflict.

53. Gen. Ambrose E. Burnside, tall, with mutton-chop sideburns, a corps commander in summer 1862, and eventual commander of the Army of the Potomac through the disastrous battle of Fredericksburg in December of that year.

54. Clearly, John and Annie's marriage had not been approved by Annie's mother, who must have thought otherwise of her daughter marrying a slave.

55. Washington may have read just enough nineteenth-century literature or war narratives to help him capture this remarkable sense of drama in a turning point in his story. He combines home, family, and war's destruction with the grace of a novelist.

56. Washington does not tell us how his grandmother, Molly, and probably aunt Maria and four children made it to Washington, D.C. That he gathered them there in the midst of war's chaos is a poignant example of the exodus of freedpeople who were trekking into the District during these very weeks.

57. Washington was slight in stature. This makes a remarkable admission in a narrative that has also just told of his soldierlike exploits in the war. His stark

ending of the narrative is difficult to explain, but among other matters, it may indicate his sense that he had seen enough of real war.

WALLACE TURNAGE, "JOURNAL OF WALLACE TURNAGE"

1. "Nurse" in this case likely means the care of other younger children.

2. Richmond slave trader and auction-house owner Hector Davis.

3. James Chalmers, Scottish-born cotton planter from Pickensville, Alabama.

4. Drove, a term usually used to describe herds or flocks of cattle or other animals; in this case a large group of slaves driven from market to points of sale.

5. Such descriptions of brutal whippings of women, stripped naked at least to the waist, are common in slave narratives. See Frederick Douglass's description of the beating of his aunt Hester in *Narrative of the Life of Frederick Douglass,* 45–46.

6. Gouging and biting were somewhat common in backwoods brawling traditions in the rural South. See Elliot J. Gorn, "'Gouge and Bite, Pull Hair and Scratch': The Social Significance of Fighting in the Southern Backcountry," *American Historical Review,* 90 (Spring 1985), 76–99.

7. Douglass also portrays the duration of his fight with slave breaker-overseer Edward Covey as two hours in duration, a small clue that Turnage may have read Douglass. See Douglass, *Narrative,* 88.

8. A structure, sometimes a small shed and sometimes quite large, housing the cotton gin.

9. Chitterlings: the small intestines of pigs, and when boiled or cooked, a common food among Southern blacks.

10. "Servants" likely means slaves in this context.

11. The major South-to-North railroad, running from Mobile on the Gulf, through Mississippi, Tennessee, and Kentucky, and into Ohio. It provided a route for Turnage's first four escape attempts.

12. A picket fence of rails for sticks.

13. Shannon Station, Mississippi.

14. Corinth had been held and fortified by the Union forces since 1862 and it had by this time a major contraband camp of ex-slaves.

15. Turnage's inclusion of such details of time and place may be indicative of

many things: A way to retrieve and control his memory; a form in which he could feel some comfort in writing; a manner in which to demonstrate the veracity of his story; and perhaps a desire, however conscious, to deflect the pain and humiliation in his experience.

16. Here, Turnage captures a central truth of why his master, Chalmers, steadfastly continued to retrieve Wallace. He was extremely valuable as a financial asset, and the idea of potential "sale" saves him from torture or murder.

17. Brooklyn, Mississippi.

18. In this remembrance, Turnage's mixed racial identity becomes very clear. It would be common in the Gulf coast region for blacks to have mixed Indian ancestry. Also, his daughter, Lydia, would eventually pass for white and claim that she was "Portugee." She had read her father's narrative.

19. A wealthy Mobile merchant, Collier Harrison Minge (1799–c.1868). Minge was from a prominent Virginia family and a nephew of former president William Henry Harrison.

20. Turnage overestimates his time with Minge by several months.

21. He may intend the word "passing" to mean the sheer density of the flow of population in the closed-off city of Mobile, rather than its racial meaning— black but passing for white. But it could have a double meaning, racial and crowd density.

22. Second attempt from Mobile.

23. Turnage came to understand the geography of the west side of Mobile Bay very well. He likely studied some maps before writing his narrative.

24. Sometimes called swamp grass.

25. Major General Gordon Granger, commander of all Union troops in the Mobile Bay region.

26. Lieutenant Junius Thomas Turner (1826–1925), Third Maryland Cavalry.

Appendix: "The Death of Our Little Johnnie"

1 Matthew 19:14. Washington's familiarity with the Bible is remarkable. This passage in Matthew includes the story of Jesus's disciples' amazement as he lays his hands on children as well as a young man, instructing them in the Ten Commandments and urging them to abandon their worldly goods. The

chapter ends with Jesus's famous declarations that "[i]t is easier for a camel to go through the eye of a needle, than for a rich man to enter into the Kingdom of God," and that "many that are first shall be last; and the last shall be first." For a man just out of slavery, and burying an infant son, these passages may have provided not only solace but a hope of redemption in the immediate aftermath of the war and emancipation.

INDEX